Lecture Notes in Bioinformatics   **11705**

Subseries of Lecture Notes in Computer Science

More information about this series at http://www.springer.com/series/5381

Milan Češka · Nicola Paoletti (Eds.)

Hybrid Systems Biology

6th International Workshop, HSB 2019
Prague, Czech Republic, April 6–7, 2019
Revised Selected Papers

Springer

Editors
Milan Češka (iD)
Brno University of Technology
Brno, Czech Republic

Nicola Paoletti
University of London
Royal Holloway
Egham, UK

ISSN 0302-9743   ISSN 1611-3349 (electronic)
Lecture Notes in Bioinformatics
ISBN 978-3-030-28041-3   ISBN 978-3-030-28042-0 (eBook)
https://doi.org/10.1007/978-3-030-28042-0

LNCS Sublibrary: SL8 – Bioinformatics

This Springer imprint is published by the registered company Springer Nature Switzerland AG
The registered company address is: Gewerbestrasse 11, 6330 Cham, Switzerland

Preface

This volume contains the papers presented at HSB 2019, the 6th International Workshop on Hybrid Systems Biology, held during April 6–7, 2019, at the Faculty of Mathematics and Physics, Charles University, Prague, Czech Republic. HSB 2019 was co-located with ETAPS 2019, the European Joint Conferences on Theory and Practice of Software, including five prime conferences and 17 satellite workshops.

HSB provides a unique forum for discussion on dynamical models in biology, with an emphasis on both hybrid systems (in the classic sense, i.e., mixed continuous/discrete/stochastic systems) and hybrid approaches that combine modelling, analysis, algorithmic and experimental techniques from different areas. HSB 2019 strengthened the focus on the design and analysis of artificial biochemical systems (e.g., engineered bacteria or molecular machines) and of medical cyber-physical systems. Hybrid systems and approaches are essential to the understanding of complex living systems, which are characterized by stochasticity and heterogeneous spatiotemporal scales. The complexity of such systems makes their formal analysis challenging, and even their simulations are often impractical, calling for appropriate model abstractions and scalable analysis methods. To overcome these challenges, HSB aims at bringing together researchers from different disciplines and at applying these methods to the study of structure, dynamics, and control mechanisms of living systems ranging from genetic regulatory networks to metabolic networks.

HSB 2019 offered a dense two-day program, including three invited talks, regular single-track sessions, and a poster session. Moreover, HSB 2019 had the honor of including a special session dedicated to the memory of Oded Maler, a very much missed member of HSB's Steering Committee and one of the founders of the workshop. The session celebrated his life and scientific contributions with three invited talks by some of his closest collaborators. All contributed talks were of high quality, and the participation was lively, interactive, and stimulating. The workshop hosted about 30 registered participants and registered a constant inflow of attendees from other co-located events at ETAPS 2019.

A highlight of HSB 2019 was the three invited talks and three talks dedicated to the memory of Oded Maler. The speakers were selected in view of the breadth and interdisciplinarity of the workshop: Marta Kwiatkowska (University of Oxford, UK, on formal methods for behavioral prediction), Michela Chiappalone (Istituto Italiano di Tecnologia, Italy, on closed-loop neuro-hybrid interfaces), Igor Schreiber (University of Chemistry and Technology of Prague, Czech Republic on stability analysis of reaction networks), Thao Dang (CNRS/VERIMAG, France, on research odyssey of Oded Maler), Alexandre Donzé (Decyphir SAS, France, on formal barbaric systems biology), and Eugene Asarin (IRIF, University Paris Diderot and CNRS, France, on timed patterns and their monitoring).

HSB 2019 had 40 Program Committee (PC) members who provided detailed reviews of the submitted contributions, out of which nine articles were accepted for presentation during the single-track sessions and appear as full or short papers in these proceedings. To ensure the highest quality for this volume, five submissions underwent a second round of review or a shepherding process before inclusion in the proceedings. The proceedings also include three invited papers from our invited speakers.

As the program co-chairs, we are extremely grateful to the PC members and the external reviewers for their work and the valuable feedback they provided to the authors. We thank all the members of the HSB Steering Committee, for their advice on organizing and running the conference. Our special thanks go to David Šafránek for helping us with the organization and for securing our sponsors. We are pleased to acknowledge the financial support kindly received from the National Center for Systems Biology of the Czech Republic (C4SYS) and the Faculty of Information Technology, Brno University of Technology, Czech Republic. We acknowledge the support of the EasyChair conference system during the reviewing process and the production of these proceedings. We also thank Springer for publishing the HSB proceedings in its *Lecture Notes in Bioinformatics* series. Finally, we would like to thank all the participants of the conference. It was the quality of their presentations and their contribution to the discussions that made the meeting a scientific success.

June 2019   Milan Češka
                                                                       Nicola Paoletti

Organization

Program Committee

Alessandro Abate	University of Oxford, UK
Frank Allgower	University of Stuttgart, Germany
Ezio Bartocci	TU Wien, Austria
Sergiy Bogomolov	Australian National University, Australia
Luca Bortolussi	University of Trieste, Italy
Giulio Caravagna	Institute of Cancer Research London, UK
Luca Cardelli	University of Oxford, UK
Milan Ceska	Brno University of Technology, Czech Republic
Elizabeth M. Cherry	Rochester Institute of Technology, USA
Michela Chiappalone	Fondazione Istituto Italiano di Tecnologia (IIT), Italy
Eugenio Cinquemani	Inria, France
Neil Dalchau	Microsoft Research, UK
Thao Dang	CNRS/VERIMAG, University of Grenoble Alpes, France
Hidde De Jong	Inria Grenoble - Rhone-Alpes, France
Tommaso Dreossi	University of California at Berkeley, USA
François Fages	Inria, Université Paris-Saclay, France
Eric Fanchon	CNRS, TIMC-IMAG, France
Jerome Feret	Inria Paris, France
Giulia Giordano	TU Delft, The Netherlands
Ádám Halász	West Virginia University, USA
Jane Hillston	University of Edinburgh, UK
Agung Julius	Rensselaer Polytechnic Institute, USA
Hillel Kugler	Bar-Ilan University, Israel
Luca Laurenti	University of Oxford
Chris Myers	University of Utah, USA
Laura Nenzi	University of Trieste, Italy
Miroslav Pajic	Duke University, USA
Nicola Paoletti	Royal Holloway, University of London, UK
Ion Petre	University of Turku, Finland and National Institute for R&D in Biological Sciences, Romania
Tatjana Petrov	University of Konstanz, Germany
Carla Piazza	University of Udine, Italy
Guido Sanguinetti	University of Edinburgh, UK
Sriram Sankaranarayanan	University of Colorado Boulder, USA
Abhyudai Singh	University of Delaware, USA
Katerina Stankova	Maastricht University, The Netherlands
James Weimer	University of Pennsylvania, USA

Verena Wolf   Saarland University, Germany
Boyan Yordanov   Microsoft Research, UK
Paolo Zuliani   Newcastle University, UK
David Šafránek   Masaryk University, Czech Republic

Invited Abstracts

Modelling and Personalisation Techniques for Behavioural Prediction and Emotion Recognition

Marta Kwiatkowska

Department of Computer Science, University of Oxford, UK

Abstract. The prevalence of wearable sensing devices and smartphones is resulting in a multitude of physiological data being collected, for example heart rate, gait and eye movement. Driven by applications in health and behavioural monitoring, as well as affective computing, there is a growing demand for computational models that are able to accurately predict multimodal features in a variety of contexts. While machine learning models excel at identifying features in physiological signals, they lack reliability guarantees and need to be adapted to the user. This talk will give an overview of modelling and personalisation techniques developed as part of the AffecTech project[1] and their applications in the context of biometric security and emotion recognition. Future challenges in this important field will also be discussed.

---

[1] http://www.cs.ox.ac.uk/projects/AFFECTech/index.html.

Timed Patterns: From Definition to Matching and Monitoring

A Survey in Memoriam Oded Maler

Eugene Asarin

IRIF, Université de Paris and CNRS, Paris, France

Abstract. At timed level of abstraction, system behaviors are considered as sequences of discrete events (from a finite alphabet) and real-valued time lapses between them; or as discrete-valued signals over continuous time. Initiated by works of Alur & Dill and aiming modeling and verification of real-time sequences, this approach became quite popular and was successfully extended to other domains. As usual in verification, sets of timed behaviors were defined by (timed) automata, and by logical formulas.

In mid 90s, Oded Maler initiated a search for simpler, suitable for engineers, and still powerful formalism to describe sets of timed behaviors. After overcoming many technical obstacles, timed regular expressions were born, and their equivalence to timed automata proven. In follow-up works, alternative formalisms have been proposed by several researchers.

In 2010s, Oded Maler and his group came back to timed regular expressions, with a new optics of pattern-matching: given a (large) record of timed behavior of a system, and a timed regular expression describing patterns of interest (e.g. faulty sequences), detect all the occurrences of the pattern in the record. Most of this research is automata-free: pattern-matching algorithms work directly on timed behaviors. Efficient algorithms have been developed and implemented, allowing off-line and on-line pattern-matching, and using several formalisms for pattern specification, and applications to monitoring prospected.

In this talk I will present the timed view on system behaviors, and the two periods of timed regular expressions: theoretical study on expressiveness from 1990s and practice-oriented works on pattern-matching and monitoring from 2010s. No special knowledge is required from the audience. This will also be a memorial talk, on Oded's philosophical, creative and personal style of choosing research topics, leading research, and supervising students and co-workers.

From Sensitive to Formal Barbaric Systems Biology

Alexandre Donzé

Decyphir SAS, France

Abstract. Oded Maler often characterized as "barbaric" some of his approaches to solving complex problems. By this, he meant the modern meaning, i.e., "unsophisticated", for example when he suggested to compute bunch of simulations to approximate reachable sets of dynamical systems - at a time when the trend was to fill pages of fancy theorems in advanced computational geometry or functional analysis. However, it is fair to say that he was in effect a true Barbarian but in the antique sense: Ancient Greeks called "Barbarians" those who were not Greek themselves. As a matter of fact, Oded as a scientist knew no boundaries: he wandered freely between theoretical computer science and applied mathematics, control theory, logics, Physics and Biology, etc. Always with humility, humor, and avid curiosity about local customs and knowledge he would bring in his own extended scientific baggage with genuine intention and efforts to mix in and contribute to the fields he was exploring.

Systems Biology was a natural target of these explorations. There he found problems related to hybrid dynamical systems, another cross-field he contributed to pioneer. Together with various collaborators including biologists, both from wet labs and theoreticians, and myself, we experimented with and improved techniques such as systematic simulation [1, 3] and the monitoring of signal temporal logic [2, 4, 5], an extension of a logic used in program verification adapted to continuous and real-world processes, to help in particular with the difficult problem of parameter uncertainty in the modeling of living systems. I will try to recount some results we obtained and the avenues of research that this work from Oded's legacy helped create and remain open today.

References

1. Donzé, A., Clermont, G., Legay, A., Langmead, C.J.: Parameter synthesis in nonlinear dynamical systems: application to systems biology. In: Batzoglou, S. (ed.) RECOMB 2009. LNCS, vol. 5541, pp. 155–169. Springer, Heidelberg (2009). https://doi.org/10.1007/978-3-642-02008-7_11
2. Donzé, A., Fanchon, E., Gattepaille, L.M., Maler, O., Tracqui, P.: Robustness analysis and behavior discrimination in enzymatic reaction networks. PloS one **6**(9), e24246 (2011)
3. Donzé, A., Maler, O.: Systematic simulation using sensitivity analysis. In: Bemporad, A., Bicchi, A., Buttazzo, G. (eds.) HSCC 2007. LNCS, vol. 4416, pp. 174–189. Springer, Heidelberg (2007). https://doi.org/10.1007/978-3-540-71493-4_16

4. Mobilia, N., Donzé, A., Moulis, J.M., Fanchon, E.: A model of the cellular iron homeostasis network using semi-formal methods for parameter space exploration. In: Proceedings First International Workshop on Hybrid Systems and Biology, HSB 2012, pp. 42–57 (2012)
5. Stoma, S., Donzé, A., Bertaux, F., Maler, O., Batt, G.: STL-based analysis of TRAIL-induced apoptosis challenges the notion of type I/type II cell line classification. PLoS Comput. Biol. **9**(5), e1003056 (2013)

Contents

Short Paper

Invited Papers

A Multimodular System to Study the Impact of a Focal Lesion in Neuronal Cell Cultures

Alberto Averna[1], Marta Carè[1], Stefano Buccelli[1,2,3],
Marianna Semprini[1], Francesco Difato[4],
and Michela Chiappalone[1(✉)]

[1] Rehab Technologies IIT-INAIL Lab, Istituto Italiano di Tecnologia,
Via Morego 30, 16163 Genoa, Italy
michela.chiappalone@iit.it
[2] Department of Informatics, Bioengineering, Robotics,
System Engineering (DIBRIS), University of Genova,
Via all'Opera Pia 13, 16145 Genoa, Italy
[3] Department of Neuroscience, Rehabilitation, Ophthalmology,
Genetics and Maternal and Child Science (DINOGMI), University of Genova,
L.go P. Daneo 3, 16132 Genoa, Italy
[4] Neuroscience and Brain Technologies, Istituto Italiano di Tecnologia,
Via Morego 30, 16163 Genoa, Italy

Abstract. Characterizing neuronal networks activity and their dynamical changes due to endogenous and exogenous causes is a key issue of computational neuroscience and constitutes a fundamental contribution towards the development of innovative intervention strategies in case of brain damage. We address this challenge by making use of a multimodular system able to confine the growth of cells on substrate-embedded microelectrode arrays to investigate the interactions between networks of neurons. We observed their spontaneous and electrically induced network activity before and after a laser cut disconnecting one of the modules from all the others. We found that laser dissection induced de-synchronized activity among different modules during spontaneous activity, and prevented the propagation of evoked responses among modules during electrical stimulation. This reproducible experimental model constitutes a test-bed for the design and development of innovative computational tools for characterizing neural damage, and of novel neuro-prostheses aimed at restoring lost neuronal functionality between distinct brain areas.

Keywords: Cell cultures · Lesion · MEA · Modularity · Spikes · Synchronization

1 Introduction

The brain is one of the most fascinating and complex organisms of the known Universe. Even if enormous advances have been done in the last decades, also thanks to the convergence of different disciplines into the study of the brain, we still know very little about how the brain works and we are not yet able to design artificial systems emulating its functionality. One of its most peculiar property is the capability to exploit

© Springer Nature Switzerland AG 2019
M. Češka and N. Paoletti (Eds.): HSB 2019, LNBI 11705, pp. 3–15, 2019.
https://doi.org/10.1007/978-3-030-28042-0_1

plasticity to allow performing cognitive or motor task even when there is a damage. This is because the brain is redundant and intrinsically modular, being composed of local networks that are embedded in networks of networks [1], sparsely connected to each other [2]: the connections can reorganize bypassing the damage or reinforcing weak connections [3, 4].

Indeed, understanding the intricacy of brain signals, what is the effect of a damage on signal generation [5], and how this impacts on the electrophysiological behavior of brain networks and on their (re)-organization, has a twofold importance: from one side it is necessary in order to shape suitable intervention strategies [6] based on novel bioelectronic devices [3]; from the other side it will help in designing novel 'neuro-biohybrid' technologies [7] which can exploit the brain self-repair capability in case of a damage. To reach the above goals it is fundamental to deeply characterize and understand how a damage affects the electrophysiological behavior of a network.

Within this framework, simplified in vitro models of cell assemblies can provide useful insights to investigate the interactions between networks of neurons, both in physiological and in pathological conditions. In vitro systems, by overcoming the limits of in vivo models imposed by the complexity of the surrounding networks and by the consequent low level of reproducibility, can thus serve as test bed for innovative solutions ranging from neuropharmacological to electroceutical applications [8–10]. Moreover, in silico models, either software or hardware, of the electrophysiological behavior of such reduced networks can be used to replace neuronal functionalities in the framework of novel neuroprosthetic devices [11, 12].

In the last decade, different groups have started to realize in vitro modular structures [13–16]. In particular, (multi) modular cell cultures plated over Micro Electrode Arrays (MEAs) represent an interesting bio-artificial experimental model for studying neuronal networks at the mesoscale level for different reasons. First, cell cultures on MEA can be manipulated in several ways (ranging from pharmacological to electrical, optical and other kinds of perturbations) and can survive longer with respect to other preparations [17]. Second, recording from multiple sites is crucial for investigating neural information processing, in case of neural networks and in particular when dealing with multimodular cell assemblies. Third, the high temporal resolution of MEAs allow characterizing the neuronal activity at a time scale that is critical to understand neuronal dynamics.

In the present work, we realized a multimodular structure able to reproduce the modular architecture of different interconnected subpopulations. We then characterized the electrophysiological activity of both spontaneous and electrically evoked activity in the obtained cell cultures confined in 3 or 4 modules. Neuronal modules, during development, projected to each other and therefore self-organized themselves in a network with intricate functional and anatomical connectivity to mimic at least a part of the modular properties of the neuronal tissue of origin. We then used a custom-made laser setup [18] able to produce a focal lesion between modules, thus affecting the anatomical connectivity among the neuronal modules. We thus observed the correlation of spike trains between modules and their changes due to the lesion.

This work lays the foundation for understanding the dynamical changes occurring after a brain lesion, a critical step towards the development of novel strategies to overcome the loss of communication between cell assemblies, with applications on both in vivo systems and in silico devices.

2 Methods

2.1 Ethics Statement

We obtained primary neuronal cultures from rat embryos at gestational day E18 (pregnant Sprague-Dawley female rats were delivered by Charles River Laboratories, Lecco, Italy). When performing the experiments, we minimized the number of sacrificed rats and the potential for nociceptor activation and pain-like sensations and we respected the three R (replacement, reduction and refinement) principle in accordance with the guidelines established by the European Community Council (Directive 2010/63/EU of September 22nd, 2010). Rat housing was in accordance with institutional guidelines and with the in force legislation of Italy (legislation N°116 of 1992). The procedures for preparing neuronal cultures are described in detail in previous studies [9, 19].

2.2 PDMS Structures for Multimodular Network Confinement

Through soft lithography, we realized a polymeric structure in polydimethylsiloxane (PDMS), composed of different modules, in order to provide the physical confinement of neuronal cultures [20, 21]. The PDMS mask was positioned on the MEA substrate before the coating procedure, performed by putting a 100-μl drop of laminin and poly-D-lysine solution on the mask and leaving it in the vacuum chamber for 20 min. The mask was then removed and cells were plated afterwards. MEAs (Multi Channel Systems, Reutlingen, Germany) are available in different geometrical layouts: "4Q", where 60 electrodes are organized in 4 separate quadrants, "8 × 8", where electrodes are placed according to a square grid; "6 × 10", where electrodes are placed according to a rectangular grid. The nominal final cell concentration was around 500 cells/μl ($\sim$ 100 cells per network module).

2.3 Laser Ablation Setup

The entire optical system was described in a previous work [18]. Briefly, the laser dissection source consisted in a pulsed sub-nanosecond UV Nd:YAG laser at 355 nm (PNV-001525-040, PowerChip nano-Pulse UV laser – Teem Photonics), whose output was modulated with the aid of an acousto-optical modulator (MQ110-A3-UV, 355 nm fused silica, AA-Opto-electronic) driven by a custom low impedance linear driver. The laser dissector was integrated in a modified upright microscope (BX51 – Olympus) equipped with a 20x, 0.5 NA water dipping objective. A custom-made software interface based on LabVIEW (National Instruments) controlled the UV laser intensity, pulse repetition rate, and the number of pulses delivered to the sample. Synchronization signals between devices were sent through a D/A board (PCI-6529, 24 bit, 4 channels, 204.8 kSamples/second, National Instruments), in order to synchronize CCD image acquisition (Andor DU-897D-C00), sample positioning trough motorized stage (assembled 3-axis linear stages, M-126.CG1, Physics-Instruments), and the trigger of UV laser pulses.

2.4   Recording Set-Up and Experimental Protocol

The activity of all cultures was recorded by means of the MEA60 System (MCS). The signal from each channel was sampled at 25 kHz and amplified using a Multichannel System amplifier with a bandwidth of 1 Hz–3 kHz. Each recorded channel was acquired through the data acquisition card and on-line monitored through MC_Rack software (MCS). A commercial stimulator (MCS) was integrated in the system to deliver electrical pulses through one of the electrodes of the MEA. Figure 1A presents a scheme of the experimental setup.

To reduce thermal stress of the cells during the experiment, MEAs were kept at 37 °C by means of a controlled thermostat (MCS) and covered by PDMS caps to avoid evaporation and prevent changes in osmolarity.

We performed experiments on neocortical modular networks, recorded between 20 and 25 Days In Vitro (DIVs). The experimental dataset consisted of 8 modular networks (Fig. 1B) consisted of 5 consecutive phases:

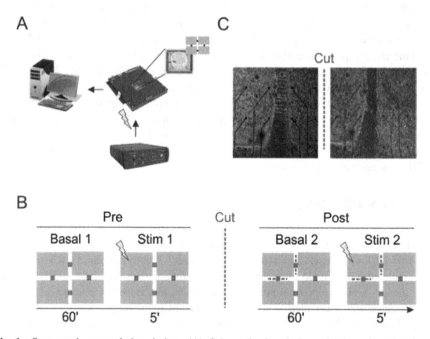

Fig. 1. Setup and protocol description. (A) Schematic description of the setup. A personal computer equipped with MC-Card (Multichannel System, MCS) records the activity from the MEA 1060 Amplifier system. A commercial stimulator (STG 4002, MCS) delivered open-loop regular stimulation to the MEA amplifier. (B) Schematic of the experimental protocol, consisting of 60 min of basal (spontaneous) activity followed by 5 min of stimulation delivered to one electrode. Stimulation was delivered to different electrodes to test the propagation of the signal across modules. One of the clusters was then isolated by means of a laser cut. Following the laser ablation, the protocol was repeated as before cut. (C) Optical micrographs depicting a corner of a modular culture before (left) and after (right) laser cut.

(i) 'Basal1': one-hour recording of spontaneous activity;

(ii) 'Stim1': stimulation session I, which consists of serially stimulating at least two electrodes per cluster using a train of 50 positive then- negative pulses (1.5 V peak-to-peak, duration 500 μs, duty cycle 50%) at 0.2 Hz;

(iii) 'Cut': laser ablation of inter-cluster neural connections, whose aim is to isolate a cluster which is physically and functionally connected to at least another one (Fig. 1C), note that in control experiments no laser ablation was performed;

(iv) 'Basal2': one-hour recording of spontaneous activity after performing the lesion;

(v) 'Stim2': stimulation session II, from the same electrodes of phase ii.

Two control experiments underwent the same protocol, but in phase (iii) the laser setup was turned off (i.e. the 'Cut' phase was not present).

2.5 Data Analysis and Statistics

Raw data were bandpass-filtered in the Multi-Unit Activity - MUA (i.e. spikes) data band (~ 300 Hz to 3 kHz) and processed using custom MATLAB (The Mathworks, Natick, MA, USA) scripts. A custom offline spike detection algorithm, based on Precise Timing Spike Detection, was used to discriminate spikes [22]. For detecting bursting activity, we used a custom Burst Detection algorithm, according to which bursts were identified as sequences of at least 5 consecutive spikes spaced less than a 100 ms time threshold [23].

Once spikes and bursts were detected, we computed the following electrophysiological parameters:

- Mean Firing Rate, MFR (mean number of spikes over an interval of time [spike/s]);
- Mean Bursting Rate, MBR (mean number of bursts over an interval of time [burst/minute]);
- Burst Duration (duration of burst in [ms])
- Burstiness Index (burstiness level of the network, providing a normalized value between 0 - no bursts - and 1 - burst dominated [9, 24]).

To characterize the effect of the lesion induced by the laser ablation, we computed the ratios of MFR, MBR and Burst Duration between their values calculated post and pre lesion.

Furthermore, we evaluated the level of pairwise correlation between the spike trains, by exploiting the Cross-Correlation method implemented in SpyCode, a custom Matlab-based software developed in our lab [25]. We measured correlations within 100 ms time windows, using 1 ms bins and we normalized the Cross-Correlation peaks found in the post lesion phase over the average value of correlation found before the lesion.

In order to investigate the impact of the lesion on the evoked activity, we calculated the Post-Stimulus Time Histograms (PSTH) of stimulus-associated action potentials detected from each electrode (1 ms bins, normalized over the total number of stimulation pulses) [26]. The area under the normalized PSTH curve was used to quantify the total amount of stimulation-evoked neural activity during each stimulation phase.

The normal distribution of experimental data was assessed using the Kolmogorov-Smirnov normality test. Statistical comparisons were performed with Wilcoxon signed-rank test, with p-values <0.05 were considered as significant.

3   Results

Typical multimodular cultures were characterized by subpopulations of neurons interconnected by neurites allowing transfer of information from one module to another (Fig. 1C, left and Fig. 2A, left). The goal of the present study was to characterize both the spontaneous and evoked dynamics of multimodular networks and to subsequently evaluate the electrophysiological effect of the laser ablation aimed to isolate one of the neuronal module (Fig. 1C, right and Fig. 2B, left).

3.1   Laser Cut Affects the Spontaneous Activity of a Neuronal Network

Spiking activity of multimodular networks appeared well synchronous for the entire duration of the experiment in control condition, as it can be qualitatively appreciated by looking at the raster plot of Fig. 2A. On the other hand, the laser dissection of a cluster induced a strong desynchronization of spiking activity between the isolated cluster and all the others (Fig. 2B).

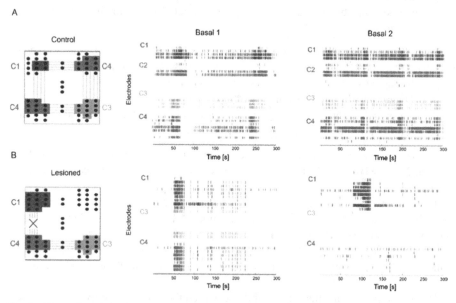

Fig. 2. Network activity in two representative experiments. (A) Control experiment. On the left MEA modules are graphically depicted, black dots representing the channels, where the active ones have background colored according to the module they belong to. In the center, a raster plot of the activity recorded during the Basal 1 phase is colored according to modules as on the left. In the right a raster plot of the activity recorded during the Basal 2 phase is colored according to modules as on the left. (B) Same as in A, for experiments with laser ablation (i.e. between Basal 1 and Basal 2, a laser ablation was performed - Cut phase). (Color figure online)

As a consequence of the lesion, we also observed a significant global decrease in the network mean firing rate, both inside the isolated cluster (Isolated, $p < 0.05$), and in all the other clusters (Others, $p < 0.001$) that were previously connected to the isolated one (Wilcoxon signed-rank test, Fig. 3A). No changes of firing rate were found in the

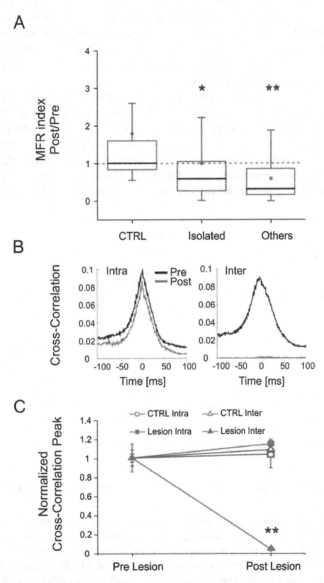

Fig. 3. Changes in spontaneous activity. (A) MFR variation in control experiments (CTRL) and in modules belonging to experiments with laser ablation (Isolated and Others). (B) Cross-Correlation pre and post lesion calculated in the same cluster (Intra) and among the isolated cluster electrodes and the other clusters (Inter) for the experiments with laser ablation. (C) Normalized cross-correlations pre and post lesion both for the lesioned (Lesion) and control (CTRL) experiments. $**p < 0.01$, $*p < 0.05$ Wilcoxon signed-rank test.

control condition (Fig. 3A). Moreover, the analysis of the pairwise correlation indicated a stable level of synchronicity for the activity of the electrodes belonging to the same cluster (Intra, Fig. 3B, left), while a dramatic drop was observed between the electrodes of the isolated cluster and all the other ones (Inter, Fig. 3B, right).

Figure 3C underlines this effect by showing the Cross-Correlation peaks normalized over the average value of correlation found before the laser dissection, calculated both in the same cluster (Intra) and among the isolated cluster electrodes and the other clusters (Inter). While correlation remained rather stable in the control and in the lesioned intra condition, it significantly dropped in the lesioned intra condition ($p < 0.001$, Wilcoxon signed-rank test).

As shown in Fig. 4, laser dissection also influenced the bursting activity (Fig. 4A). We found that, after ablation, the tendency of both the rate (Fig. 4B) and the duration (Fig. 4C) of the detected bursting activity was to decrease for all the recorded clusters in the lesioned condition, while they remained stable for the control condition (values close to 1). Moreover the burst duration significantly decreased after ablation ($p < 0.001$, Wilcoxon signed-rank test). The level of burstiness of the network showed a tendency to increase within the isolated cluster after the lesion (Fig. 4D).

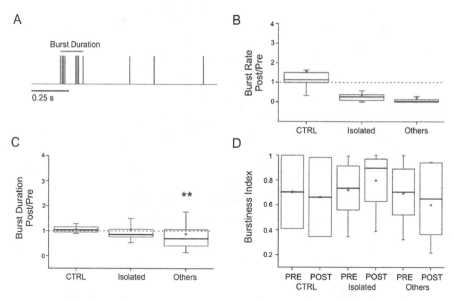

Fig. 4. Comparison of spontaneous and evoked bursting activity of healthy and damaged modular cultures. (A) 1.25-s raster plot of spontaneous activity of a channel before lesion. The red lines correspond to the detected bursting activity, the red bar above them represents the Burst Duration, and the grey lines corresponds to isolated spiking activity. (B) Variation of Burst Rate pre and post lesion either within the isolated cluster (Isolated) or in the other ones (Others) and in the no-lesioned ones (CTRL). (C) Comparison between the statistical distributions of the ratio between normalized Burst Durations calculated pre and post lesion for the isolated cluster (Isolated), the other clusters (Others) and for the no-lesioned (CTRL) condition. (D) Burstiness Index calculated pre and post lesion in all condition. **$p < 0.01$, Wilcoxon signed-rank test. (Color figure online)

3.2 Laser Dissection Confines Evoked Activity

We also observed the effect of electrical stimulation in the different conditions. Before performing the lesion, electrical stimulation (Fig. 5A) was able to evoke activity both within the cluster hosting the stimulation channel and in the connected modules (Fig. 5B). After laser dissection, the evoked activity remained confined within the isolated cluster without spreading towards the other ones (Fig. 5C).

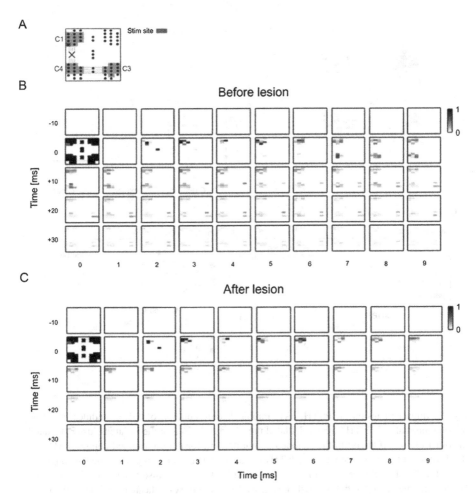

Fig. 5. Network response to stimulation. (A) Graphical representation of the stimulation site over the MEA. Stimulation was a sequence of 50 stimuli delivered at 0.2 Hz through an electrode in the top left module before lesion. (B) Stimulation effects in the pre lesion condition. Each graph represents the 60-electrodes MEA response at different time points with respect to the stimulus: ranging from −10 ms before the stimulation to 30 ms after the stimulus onset. Each pixel represents, in grey level, the probability to detect a spike in 1 ms bin. Around the stimulus onset, the spike detection algorithm identifies artifacts on all electrodes that were then blanked to avoid false positive detections. Stimulation site is highlighted in red. (C) Stimulation effects in the post lesion condition. Same as in A, but after lesion. (Color figure online)

We quantified this effect in terms of PSTH area variation (Fig. 6). We observed a significant (Wilcoxon signed-rank test, p < 0.001) global decrease of PSTH Area after the laser cut in both local and distal responses (Fig. 6B): since lesion was aimed at reducing the amount of connections among clusters, responses of other clusters resulted to be more affected than that the isolated cluster. Single channel PSTH variation are highlighted in Fig. 6C.

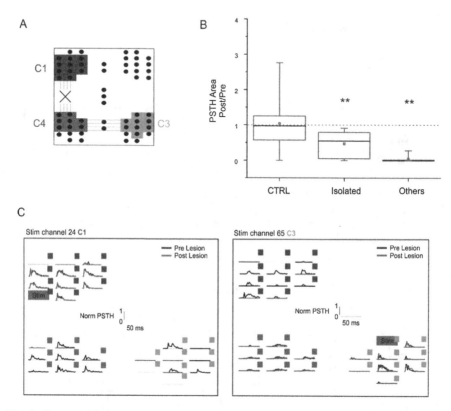

Fig. 6. Impact of lesion on PSTH. (A) Graphical representation of MEA modules. (B) Ratio between normalized PSTH Areas calculated pre and post lesion for the isolated cluster (Isolated), the other clusters (Others) in the lesioned condition, and for the no-lesioned condition (CTRL). (C) Post-stimulus time histograms (PSTH) obtained when stimulating a channel belonging to either the isolated cluster (left) or one channel from the other clusters (right). The electrodes labelled in blue belong to the isolated cluster, while the red boxes indicate the stimulated channel. Black curves report the evoked activity before the lesion, while the red ones refer to the evoked activity after lesion. **p < 0.01, *p < 0.05 Wilcoxon signed-rank test. (Color figure online)

4 Discussion

Despite their simplicity, multimodular cell cultures are a very useful tool to manipulate the neuronal networks dynamics. Modular networks were recently developed to impose a predefined directionality in functional information transfer between neighboring nodes [14]. A modular preparation was also recently used to prove the importance of the modular organization on dynamical richness in cortical networks [16].

In this work, we characterized the electrophysiological behavior of multimodular networks coupled to MEAs, devices able to both record and stimulate the neuronal activity of the neuronal cells sitting on the surface of the planar electrodes. Thanks to this system, we were able to show that selective laser dissection of interconnections among neural assemblies affected both spontaneous and evoked activity of multimodular networks, by inducing de-synchronization between the different modules during spontaneous activity, and preventing propagation of evoked responses among modules. The activity remained confined to the isolated cluster exhibiting the same level of synchronization as before the lesion. Moreover, the burstiness level showed a tendency to increase after the lesion, suggesting that the lack of incoming/outcoming connections further promoted the appearance of network-wide events. This is in line with recent findings related to the concept of 'dynamical richness' and 'network complexity' [16, 27], according to which interconnected subpopulation of neurons show a richer dynamics than single isolated clusters of cells.

Controlling the parameters of network information and studying the effect of a lesion can have a disruptive impact both in the neuroscience and in the computational neuroscience field. One of the open questions in neuroscience concerns our ability to decode the electrophysiological patterns of network activity [28]. The use of engineered neuronal networks can be the key to further investigate the neural code. Moreover, computational neuroscience can benefit from studies like ours in which simple and controllable biological networks are involved, to better tune the computational models that aims at mimicking the electrophysiological activity of the brain [29]. This kind of studies also opens up new avenues towards 'wetware' based technologies which can be employed in a synergistic way with pure silicon-based systems to truly emulate brain's activity and reproduce intelligent behaviors [30].

In the coming years it will be increasingly necessary to take advantage of the complementary strengths of biological and computational studies to face the increasingly complex challenges related to study of the brain. All this will then promote interesting opportunities for innovative technologies to treat brain-related disabilities or deriving breakthrough 'neuro-computational' methodologies.

Acknowledgments. The presented research results has received funding from the European Union's Seventh Framework Programme (ICT-FET FP7/2007-2013, FET Young Explorers scheme) under grant agreement n° 284772 BRAIN BOW (www.brainbowproject.eu).

The authors are grateful to Dr. Valentina Pasquale for the supervision of Dr. Averna for data analyses and for useful discussions on the final results, to Dr. Paolo Bonifazi and Dr. Marta Bisio for the design of the multimodular cultures and for performing the first experiments. The authors would also like to thank Dr. Marina Nanni, Dr. Claudia Chiabrera, and Dr. Giacomo Pruzzo for precious technical support in performing the in vitro experiments at IIT.

References

1. Meunier, D., Lambiotte, R., Fornito, A., Ersche, K.D., Bullmore, E.T.: Hierarchical modularity in human brain functional networks. Front. Neuroinform. **3**, 37 (2009). https://doi.org/10.3389/neuro.11.037.2009
2. Levy, O., Ziv, N.E., Marom, S.: Enhancement of neural representation capacity by modular architecture in networks of cortical neurons. Eur. J. Neurosci. **35**, 1753–1760 (2012). https://doi.org/10.1111/j.1460-9568.2012.08094.x
3. Guggenmos, D.J., et al.: Restoration of function after brain damage using a neural prosthesis. Proc. Natl. Acad. Sci. USA **110**, 21177–21182 (2013). https://doi.org/10.1073/pnas.1316885110
4. Nudo, R.J.: Recovery after damage to motor cortical areas. Curr. Opin. Neurobiol. **9**, 740–747 (1999)
5. Bassett, D.S., Bullmore, E.T.: Human brain networks in health and disease. Curr. Opin. Neurol. **22**, 340–347 (2009). https://doi.org/10.1097/WCO.0b013e32832d93dd
6. Kleim, J.A., Jones, T.A.: Principles of experience-dependent neural plasticity: implications for rehabilitation after brain damage. J. Speech Lang. Hear. Res. **51**, S225–S239 (2008). https://doi.org/10.1044/1092-4388(2008/018)
7. Vassanelli, S., Mahmud, M.: Trends and challenges in neuroengineering: toward "intelligent" neuroprostheses through brain-"brain inspired systems" communication. Front. Neurosci. **10**, 438 (2016). https://doi.org/10.3389/fnins.2016.00438
8. Bonifazi, P., et al.: In vitro large-scale experimental and theoretical studies for the realization of bi-directional brain-prostheses. Front. Neural Circ. **7**, 40 (2013). https://doi.org/10.3389/fncir.2013.00040
9. Colombi, I., Mahajani, S., Frega, M., Gasparini, L., Chiappalone, M.: Effects of antiepileptic drugs on hippocampal neurons coupled to micro-electrode arrays. Front. Neuroeng. **6**, 10 (2013). https://doi.org/10.3389/fneng.2013.00010
10. Tessadori, J., Bisio, M., Martinoia, S., Chiappalone, M.: Modular neuronal assemblies embodied in a closed-loop environment: toward future integration of brains and machines. Front. Neural Circ. **6**, 99 (2012). https://doi.org/10.3389/fncir.2012.00099
11. Buccelli, S., et al.: A neuroprosthetic system to restore neuronal communication in modular networks. bioRxiv, 514836 https://doi.org/10.1101/514836 (2019)
12. Berger, T.W., et al.: A hippocampal cognitive prosthesis: multi-input, multi-output nonlinear modeling and VLSI implementation. IEEE Trans. Neural Syst. Rehabil. Eng. **20**, 198–211 (2012). https://doi.org/10.1109/TNSRE.2012.2189133
13. Bisio, M., Bosca, A., Pasquale, V., Berdondini, L., Chiappalone, M.: Emergence of bursting activity in connected neuronal sub-populations. PLoS ONE **9**, e107400 (2014). https://doi.org/10.1371/journal.pone.0107400
14. Forro, C., et al.: Modular microstructure design to build neuronal networks of defined functional connectivity. Biosens. Bioelectron. **122**, 75–87 (2018). https://doi.org/10.1016/j.bios.2018.08.075
15. Shein-Idelson, M., Ben-Jacob, E., Hanein, Y.: Engineered neuronal circuits: a new platform for studying the role of modular topology. Front. Neuroeng. **4**, 10 (2011). https://doi.org/10.3389/fneng.2011.00010
16. Yamamoto, H., et al.: Impact of modular organization on dynamical richness in cortical networks. Sci Adv. **4**, eaau4914 (2018). https://doi.org/10.1126/sciadv.aau4914
17. Potter, S.M., DeMarse, T.B.: A new approach to neural cell culture for long-term studies. J. Neurosci. Methods **110**, 17–24 (2001)

18. Difato, F., et al.: Combined optical tweezers and laser dissector for controlled ablation of functional connections in neural networks. J. Biomed. Opt. **16**, 051306 (2011). https://doi.org/10.1117/1.3560268

19. Frega, M., et al.: Cortical cultures coupled to micro-electrode arrays: a novel approach to perform in vitro excitotoxicity testing. Neurotoxicol. Teratol. **34**, 116–127 (2012). https://doi.org/10.1016/j.ntt.2011.08.001

20. Shein Idelson, M., Ben-Jacob, E., Hanein, Y.: Innate synchronous oscillations in freely-organized small neuronal circuits. PLoS ONE **5**, e14443 (2011). https://doi.org/10.1371/journal.pone.0014443

21. Kanner, S., et al.: Design, surface treatment, cellular plating, and culturing of modular neuronal networks composed of functionally inter-connected circuits. J. Vis. Exp. (2015). https://doi.org/10.3791/52572

22. Maccione, A., et al.: A novel algorithm for precise identification of spikes in extracellularly recorded neuronal signals. J. Neurosci. Methods **177**, 241–249 (2009). https://doi.org/10.1016/j.jneumeth.2008.09.026

23. Chiappalone, M., et al.: Burst detection algorithms for the analysis of spatio-temporal patterns in cortical networks of neurons. Neurocomputing **65–66**, 653–662 (2005). https://doi.org/10.1016/j.neucom.2004.10.094

24. Wagenaar, D.A., Madhavan, R., Pine, J., Potter, S.M.: Controlling bursting in cortical cultures with closed-loop multi-electrode stimulation. J. Neurosci. **25**, 680–688 (2005). https://doi.org/10.1523/JNEUROSCI.4209-04.2005

25. Bologna, L.L., et al.: Investigating neuronal activity by SPYCODE multi-channel data analyzer. Neural Netw. **23**, 685–697 (2010). https://doi.org/10.1016/j.neunet.2010.05.002

26. Rieke, F., Warland, D., Steveninck, R.d.R.v., Bialek, W.: Spikes: Exploring the Neural Code. MIT Press, Cambridge (1999)

27. Casali, A.G., et al.: A theoretically based index of consciousness independent of sensory processing and behavior. Sci. Transl. Med. **5**, 198ra105 (2013). https://doi.org/10.1126/scitranslmed.3006294

28. Panzeri, S., Harvey, C.D., Piasini, E., Latham, P.E., Fellin, T.: Cracking the neural code for sensory perception by combining statistics, intervention, and behavior. Neuron **93**, 491–507 (2017). https://doi.org/10.1016/j.neuron.2016.12.036

29. Maass, W., Natschläger, T., Markram, H.: 575-605-575-605, Chapman & Hall/CRC, Hoboken

30. Brooks, R., Hassabis, D., Bray, D., Shashua, A.: Turing centenary: is the brain a good model for machine intelligence? Nature **482**, 462–463 (2012). https://doi.org/10.1038/482462a

Reachability Analysis and Hybrid Systems Biology - In Memoriam Oded Maler

Thao Dang[✉]

VERIMAG/CNRS, Université Grenoble Alpes, Grenoble, France
thao.dang@univ-grenoble-alpes.fr

Abstract. In this note we present some influential contributions of Oded Maler in hybrid systems research, with a focus on his pioneering results in reachability analysis and applications to systems biology. We also give a brief discussion of the evolution of the reachability computation techniques which have greatly progressed in recent years. This discussion is not intended to include an exhaustive survey of the existing results (The reader is referred to the recent proceedings of the conferences Hybrid Systems: Computation and Control.) but to show the strong impact of his foundational work.

1 Modelling and Decidability Results

The years 80s witnessed a growing interest in timed systems that combine discrete models with metric time, in order to specify behaviours of reactive systems not only qualitatively but also quantitatively. This interest, which remains vibrant today, led to the development of a variety of formal models and logics. Timed automata [5], introduced together with a verification algorithm in the early 90s by Rajeev Alur and David Dill, have been undoubtedly the most popular formalism. They are used in many successful tools, such as UPPAAL [34], for specifying and verifying real-time systems. In the 80s, Oded was a PhD student at Weizmann Institute of Science, working on his thesis titled "Finite Automata: Infinite Behavior, Learnability and Decomposition", under the supervision of Amir Pnueli. His advisor, winner of the Turing award in 1996 for his work on temporal logics, was part of the timed systems movement. He proposed a model, called *timed transition system*, and versions of real-time temporal logics [49]. This activity of his advisor certainly had a lot of impact on Oded who was already interested in the physical world outside computers and programs. He discovered a paper by R. Brooks from MIT AI lab which proposed a "behaviour-based" approach to robotic systems integrating control programs, sensors, actuators and timers. This prompted him to think how one can verify that such systems behave correctly in a given environment. Together with Amir Pnueli, he wrote a proposal titled "Systematic Development of Robots". This idea perhaps sounded too avant-garde at that time, partly because of the inter-cultural gap between computer scientists and control theorists. In addition, the verification

M. Češka and N. Paoletti (Eds.): HSB 2019, LNBI 11705, pp. 16–29, 2019.
https://doi.org/10.1007/978-3-030-28042-0_2

research did not yet reach its successes in industrial applications. The proposal did not pass and Oded moved to France to do his postdoc at IRISA (Rennes). During his post-doc, together with Amir Pnueli and Zohar Manna, he wrote the paper "From timed to hybrid systems" [59] which proposed a model called *phase transition systems*, the first formal hybrid model coming from the verification community. This model, which combines discrete transitions (that take no time) and continuous dynamics specified by differential equations, can be thought of as a precursor of the model of hybrid automata proposed a little later in a seminal paper [3]. In parallel, various models were proposed by the control community. These models, designed to include specific discrete phenomena arising in computer control systems (such as autonomous or controlled switchings and jumps), are suitable for the purpose of extending the existing analytic control methods to hybrid systems. Many studies were devoted to the topology and computational capabilities of such models. The reader is referred to the PhD thesis of Michael S. Branicky at MIT in 1995 [26] for a thorough survey of the hybrid models proposed in the beginning of the hybrid systems research history, including a technical comparison and classification of these models. Many of these models (such as, hybrid automata and switched systems) are now widely used for verification and control purposes. These models were however too complex for the verification algorithms developed by computer scientists, who still considered differential equations outside the traditional scope of their domain. Motivated by the success of decidability results and model-checking algorithms for timed automata, the verification community was then more interested in extending these results to hybrid automata with simple continuous dynamics, such as with clocks that can be stopped, or with continuous variables having constant or piece-wise constant derivatives [4,48]. Oded was by that time an CNRS researcher in the laboratory VERIMAG, headed by Joseph Sifakis whose group contributed to the development of the hybrid automaton model [3]. In this new movement, Oded's contributions were the decidability and undecidability results for piecewise constant derivative systems (PCD). Such a system consists of a polyhedral partition of the state space and in each region of the partition the continuous variables evolve with constant derivatives. He first proved decidability of such systems with 2 continuous variables (planar PCD) [61], based on the observation that a trajectory cannot intersect itself (Jordan curve theorem), unlike the trajectory depicted in Fig. 1. Additionally, for every trajectory, the sequence of edges it crosses is ultimately-periodic. Therefore, one can define a finite abstract alphabet to describe qualitative behaviours as sequences of regions or edges. Then, Oded and his colleague Eugene Asarin proved the undecidability of PCD in 3 and higher dimensions [12,14]. They also proved, using Zeno paradox, how all the arithmetical hierarchy can be realized by PCD [15]. These results and directions have a number of important technical follow-ups, among which we can mention: a generalisation to planar differential inclusions [17], stability of polyhedral switched systems [68], and in particular models of computation [24]. Despite the recognition these theoretical results received, Oded was disappointed because these negative decidability results seemed to imply

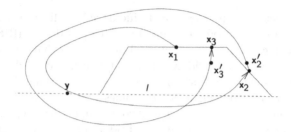

Fig. 1. Illustration of an impossible situation: the depicted trajectory exits a region at a boundary point x_1 and then exits this region again at a boundary point x_2, thus it cannot intersect the boundary part between x_1 and x_2.

that if the verification problem for such simple systems is undecidable, there would be no hope that one can verify real-life hybrid systems. This disappointment reflected his ambition to export the verification methodology to practical application domains. It also made him question the appropriateness of the exact formulation of verification problems in the context of hybrid systems. While pondering upon this methodological issue, Oded continued to work actively on timed systems. Indeed he never stopped working on this theme and his contributions in timed systems were abundant and impactful, covering a large number of problems: controller synthesis for timed automata [16,62], scheduling using timed automata (with optimality and under stochastic uncertainty) [1,52], compositional timing analysis [69], control with bounded computational resources [58], multi-criteria optimisation [30], embedded multicore [71], timed regular expressions [13], real-time temporal logic, monitoring, timed pattern matching [20]. It is important to emphasise that his results on Signal Temporal Logic [60] not only successfully gained industrial acceptance but also opened new research directions in cyber-physical systems monitoring and testing (see for example [21]).

2   Reachability Analysis

While negative decidability results hindered a direct application of the algorithmic verification methodology to hybrid systems with non-trivial dynamics, they also incited the verification community with a new motivation. In a continuous world, it is meaningful to seek approximate answers for non-trivial systems, rather than insisting on exact answers which are possible only for trivial systems. Oded set out to tackle the first obstacle: continuous systems described by differential equations. While restricting to the problem of approximating reachable sets for this type of dynamics, he aimed at a solution that could be extended to hybrid systems and could be used further for problems beyond verification, in particular for controller synthesis. Although the type of dynamics was restricted, the goal turned out to be very ambitious because the considered class is general, including non-linear dynamics. With his student, he developed a method for tracking the evolution of a (general) polyhedron under continuous dynamics.

Essentially, due to continuity of trajectories, it suffices to track the evolution of the faces of the polyhedron. A face is pushed outward if there is a point on the face at which the projection of the vector field on the normal of the face points outward, and the pushing amount depends on the time step size and the maximum projection magnitude over all the points on the face. This results in a new polyhedron for which the same procedure is applied in the next time step. This method can be seen as a set-valued Euler integration scheme. To ensure accurate results, the faces with a large derivative variation need to be subdivided, which generates non-convex polyhedra. The development of this method was much inspired by the work of Greenstreet [44] in 1996 where this idea was proposed for two dimensional systems and reachable sets are thus polygons. Non-convex polygons benefit from well-developed plane geometric manipulation algorithms, unlike general dimensional non-convex polyhedra. Treating high dimensional systems was never seen as ambitious, since real-life models are rarely limited to a few variables. It was thus necessary to choose a set representation on which the specific operations (such as pushing, splitting) as well as the Boolean set-theoretic operations (intersection and union, for handling discrete transitions) can be efficiently computed. To this end, orthogonal polyhedra (which are union of hyper-rectangles) were used [25]. The method, called "face-lifting" (see Fig. 2), was published in the proceedings of the conference HSCC (Hybrid Systems: Computation and Control) 1998 [32]. Although disappointment ensued again when it became clear that the face-lifting technique was very computationally expensive, this paper turned out to be well received by the hybrid systems community and obtained a test-of-time award at the conference HSCC 2019, to the surprise of the (living) co-author (and, plausibly, of the other co-author too). The reason was perhaps that this paper, by stating a reachability analysis problem and describing the ingredients necessary for designing an effective algorithm to solve it, opened a new concrete direction for hybrid systems verification. Indeed, as attested by the publications at the HSCC conferences, reachability analysis itself has become a central problem.

The experience with non-linear systems made Oded and his student more aware of the importance of exploiting the structure of the system. Together with Eugene Asarin, they focused on linear systems, for which the reachable set can be constructed from a finite number of trajectories via the convex hull operations. This allowed them to obtain a second-order approximation scheme that uses convex polyhedra to represent reachable sets [7,8]. The method was then extended to linear systems with uncertain input using the Maximum Principle from optimal control. These results were implemented in the tool d/dt [11]. The extension to uncertain input was inspired by the ellipsoidal technique of Kurzhanski [54] and the polyhedral technique of Varaiya [74], developed in the context of uncertain systems (although these techniques worked only for discrete-time reachable sets). Another related work was that by Chutinan and Krogh [28], who proposed a similar polyhedral approximation for systems with constant input. Besides the direct ordinary differential equation (ODE) formulation, the reachability problem was tackled using the partial differential equation

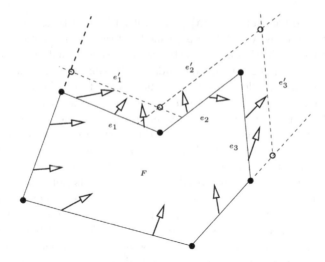

Fig. 2. Illustration of the face-lifting technique.

(PDE) formulation [63,64,73] and level sets were used to represent reachable sets. While computer scientists tried to handle differential equations, control theorists became interested in the decidability question and contributed fundamental results for hybrid systems with linear continuous dynamics [4,6,57].

In the quest for efficient set representations to make reachability algorithms more scalable, Greenstreet and Mitchell extended their method to polygonal projections [42,43]. Antoine Girard proposed zonotopes, for which computing linear transformation and Minkowski sum can be done in an algebraic manner, and this allows tracking the evolution of zonotopes under linear dynamics efficiently, without resorting to expensive geometric computations (in particular the convex hull operations) [39]. The algebraic manipulation was later adapted to general convex sets represented by support functions, in the thesis work of Colas Le Guernic, supervised by Oded and Antoine [40,41,45,46]. This thesis work culminated in a method that could compute reachable sets for systems of hundreds of dimensions. Special attention also put on performing numerical schemes intelligently to avoid error accumulation. The representations by zonotopes and support functions were implemented in the tool SpaceEx [38], developed by Oded's group under the direction of Goran Frehse. The tool quickly became one of the most advanced tools for hybrid systems verification. The influence of the work on zonotope-based reachability computation was attested by the HSCC 2018 test-of-time award given to Antoine's first paper on this topic [39].

It is fair to say that using these set representations the reachability problem for linear continuous systems is solved; however computing intersection of their unions remains (until now) a big challenge. This is a reason why the state-of-the-art reachability analysis techniques can handle purely continuous systems of up to billions of dimensions [19] but are still limited when handling hybrid systems (especially with a large number of discrete transitions).

3   Non-linear Systems and Hybrid Systems Biology

The research directions in systems biology that Oded pursued involved building and analysing dynamical system models of biological phenomena. For engineering systems, this approach is termed model-based, in the sense that a model is developed and used to debug, since correcting a model is cheaper than fixing a real system. Similarly, testing and exploring biological models in silico is preferred over expensive experiments. A network of interacting genes and proteins is thinkable as an information processing system that evolves in space and time according to fundamental laws of physics, and can thus be formally described in mathematical terms. Therefore, intuitively speaking, the biological modelling activity consists of discovering a dynamical model that can explain the relation between a diagram of biological interactions and experimental data obtained by measuring some entities in the diagram.

Whereas hybrid systems became a mathematical model widely accepted for reasoning about interactions between discrete and analog parts of embedded and cyber-physical systems, they also drew a lot of attention of researchers in systems biology since they can capture phenomena of hybrid nature in molecular biology. Oded was one of the founders of the workshop Hybrid Systems Biology. The term "hybrid systems biology" can be understood (literatim) as a branch of systems biology which relies on the techniques developed in the domain of hybrid systems. In a more allegorical manner, this term expresses a view of thinking and reasoning about biological mechanisms and processes in the spirit of the mathematical and computational methods for specifying and analysing behaviours of heterogenous systems with mixed discrete-continuous dynamics. It is important to emphasise before continuing that this note focuses only on the synergy between Oded's reachability analysis research and his interest in systems biology. He also approached systems biology via his research on real-time temporal logics for specifying and testing biological hypotheses, such as [35, 66, 70].

Oded created collaborations with some biologists (having a reciprocal interest) in Grenoble, in an effort to apply hybrid systems verification technology to biological systems, in particular the techniques that can be used to analyse in a systematic manner quantitative models admitting uncertainty whose nature is set-theoretic. Parameter uncertainty in biological models is uncertainty of this type. These collaborations led to the following observation. Hybrid systems can be used not only as a model but also to approximate complex systems by simpler ones (which can be analysed by more efficient techniques). In addition, they can naturally capture stiffness in continuous dynamics arising in many biological systems, which often causes instability in traditional numerical methods. However, their use does not come for free. Indeed, even when continuous dynamics can be efficiently handled (such as linear dynamics), discrete dynamics (which in principle can be handled using well-developed techniques for discrete systems) may lead to significant computation effort, as costly as that for overcoming numerical instability. Indeed, while numerical instability can be addressed by reducing the time step in order to adapt to fast changes of some variables, switching continuous dynamics via discrete transitions in a hybrid system may deteriorate "nice" geometric structures of continuous reachable sets. As an example, trajectories starting from points in a convex polyhedron can reach a transition guard at very different times, and the accumulation of starting points for the next continuous dynamics may form a "curved" non-convex set with complex geometry.

Taking both the drawbacks and advantages of the hybrid systems methodology into account, Oded's group revisited the hybridization approach developed in [9,10]. The main idea of this approach is to decompose a non-linear vector field into different segments corresponding to disjoint regions of the state space. Each segment is then approximated with a simpler (such as linear) vector field. This approach is very general and in principle can be applied to a large class of non-linear systems. However, in practice, the price for having artificial discrete transitions is often high, since the simpler approximate dynamics are used, the larger number of segments is needed in order to assure a desired precision. One way to avoid intersections with the boundary of two adjacent regions is to "smoothen" the transitions without compromising the approximation quality. Furthermore, geometric properties of the dynamics should be exploited to determine approximation domains that are as large as possible. This work shows an interplay between ideas from geometric modelling and set-based numerical integration, which is sketched in the following.

Given a non-linear system $\dot{x}(t) = f(x(t))$, $x \in \mathcal{X} \subset \mathbb{R}^n$ where the function f is Lipschitz. One can approximate this original system with a system: $\dot{x}(t) = g(x(t)) + u(t)$, $x \in \mathcal{X}$. The input $u(\cdot)$ such that $||u(\cdot)|| \leq \mu$ where μ is the bound of $||g - f||$, is added in the approximate system in order to conservatively account for the dynamics approximation error. The construction of such an approximate system consists of two main steps. Inside a zone of interest that contains the current reachable set, an approximation domain and its associated approximate vector field are computed. When the system leaves the current approximation domain, a new domain is created. This technique was implemented using linear interpolation over simplicial domains and multi-affine interpolation over hyper-rectangles (the interpolants in both cases can be uniquely determined). Note that the error in the reachable set approximation depends on the dynamics error bound μ. It is thus important to derive tight error bounds. For systems satisfying some smooth conditions, [33] proved for each simplex an error bound that depends on the maximal curvature of f in the simplex and on the radius of the smallest ball that contains the simplex. This error bound is tighter than the error bound used in [10] which depends on the maximal simplex edge length. In addition, one can obtain a larger simplex by stretching an equilateral simplex along a direction in which the curvature is small. This can be done by mapping the simplices to an "isotropic" space where the curvature bounds are isotropic. An illustration of this transformation is depicted in Fig. 3, where the application of the mapping to an ellipsoid produces a circle. When applying the mapping to the triangle inscribed the ellipsoid shown on the left, the result is a more regular triangle shown on the right. This mapping can be used further to define optimal shape and orientation of the simplicial domains. This dynamic hybridization based on dynamics curvature allowed treating a number of biological systems with up to 12 continuous variables [31]. This constituted a considerable progress since the original hybridization approach was limited to systems with only 3, 4 continuous variables. On the other hand, most of the existing state-of-the-art techniques for non-linear systems worked efficiently only for low-dimensional systems.

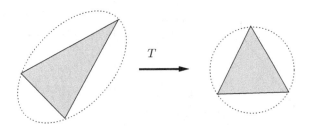

Fig. 3. Illustration of the transformation to the isotropic space.

In the following, we illustrate the results obtained by this approach for a model describing the loosening of the extra-cellular matrix [51]. This is a crucial process in angiogenesis, the sprouting of new blood vessels as a reaction to signals that indicate the need for additional oxygen in certain tissues [51,75]. Interfering with angiogenesis is considered a promising direction for fighting cancer tumors by cutting their blood supply. The soluble and membrane-associated matrix metalloproteinases are among the enzymes responsible for the proteolytic processes that occur in the extra-cellular matrix. In [51], a network of reactions involving the entities of interest was established, and then, from this network, a system of ordinary differential equations was derived using mass action kinetics. This differential equation system of 12 variables, used in our reachability analysis case study (see Table 3 of [31]), can be used to describe the proteolysis of collagen I by matrix metalloproteinases 2 (MMP2) and membrane type 1 matrix metalloproteinases (MT1-MMP) in the presence of the tissue inhibitor of metalloproteinases 2 (TIMP2). The model focuses on the degradation of collagen type 1 (represented by the variable $c1$) by two enzymes MT1-MMP and MMP2 (the concentrations of which are represented by the variables $mt1$ and $m2$). The latter has to be activated from its passive form M2P obtained by a chain of reactions involving TIMP2 (the concentration of which is represented by the variable $t2$) which also plays the role of an inhibitor for MT1-MMP, which leads to an overall complex system of interactions. The study in [51] experimentally observed a convergence of the variables, stating from a single initial state of concentrations, towards a nearly steady state (see Fig. 2-A in [51]). We computed reachable sets to verify this observation for a set of initial concentrations. Figure 4 shows the projection of the reachable set evolution on the three variables $mt1$, $m2$, and $t2$. The initial set is a small set around the origin (corresponding to the cube in the figure). We observe that the variables converge towards the dense part of the reachable set (drawn in cyan colour). This confirmed the observation of convergence in [51].

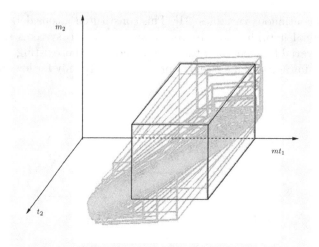

Fig. 4. Projection of the reachable set on the first three variables $mt1$, $m2$ and $t2$. The cube in this figure is the set of initial concentrations. We can observe that the variables converge towards the dense part of the reachable set (drawn in cyan colour), which confirmed the experimental observation of convergence in [51] (Color figure online).

Oded once said humorously, "our computational methods are not fast enough [to fight cancer]", referring to his ongoing Plan Cancer project (MoDyLAM - Dynamic modelling of iron-linked redox perturbations in Acute Myeloid Leukemia). His adventure in hybrid systems biology was not long, but he already paved a research path for us to follow.

4   Conclusion

This note was written in memoriam Oded Maler, who made groundbreaking contributions in the hybrid systems research. His creativity, courage, sharp mind and passion made him a role model to many of his colleagues. To celebrate Oded's scientific legacy, nothing would be more cheerful than a list[1] of major hybrid verification tools which have been developed over the last two decades: Coho [43], CheckMate [28], HyperTech [47], MPT [56], HJB toolbox [64], ET Toolbox [55], KeYmaera [65], SpaceEx [38], Adriane [29], HySon [23], NLToolbox [72], Flow* [27], CORA [2], dReach [53], C2E2 [37], AVERIST [67], HyReach [50], Sapo [36], HyLaa [18], and JuliaReach [22].

Acknowledgements. This note would not exist without the author's numerous exchanges with Oded Maler and Eugene Asarin over the last two decades. Many details about Oded's early career come from his Habilitation thesis and his various scientific writings.

---

[1] Which may be non-exhaustive.

References

1. Abdeddaïm, Y., Asarin, E., Maler, O.: Scheduling with timed automata. Theor. Comput. Sci. **354**(2), 272–300 (2006)
2. Althoff, M., Grebenyuk, D., Kochdumper, N.: Implementation of Taylor models in CORA 2018. In: Proceedings of the 5th International Workshop on Applied Verification for Continuous and Hybrid Systems, pp. 145–173 (2018)
3. Alur, R., et al.: The algorithmic analysis of hybrid systems. Theor. Comput. Sci. **138**(1), 3–34 (1995)
4. Alur, R., Henzinger, T.A., Lafferriere, G., Pappas, G.: Discrete abstractions of hybrid systems. Proc. IEEE **88**, 971–984 (2000)
5. Alur, R., Dill, D.: The theory of timed automata. In: de Bakker, J.W., Huizing, C., de Roever, W.P., Rozenberg, G. (eds.) REX 1991. LNCS, vol. 600, pp. 45–73. Springer, Heidelberg (1992). https://doi.org/10.1007/BFb0031987
6. Anai, H., Weispfenning, V.: Reach set computations using real quantifier elimination. In: Di Benedetto, M.D., Sangiovanni-Vincentelli, A. (eds.) HSCC 2001. LNCS, vol. 2034, pp. 63–76. Springer, Heidelberg (2001). https://doi.org/10.1007/3-540-45351-2_9
7. Asarin, E., Bournez, O., Dang, T., Maler, O.: Approximate reachability analysis of piecewise-linear dynamical systems. In: Lynch, N., Krogh, B.H. (eds.) HSCC 2000. LNCS, vol. 1790, pp. 20–31. Springer, Heidelberg (2000). https://doi.org/10.1007/3-540-46430-1_6
8. Asarin, E., Bournez, O., Dang, T., Maler, O., Pnueli, A.: Effective synthesis of switching controllers for linear systems. Proc. IEEE **88**, 1011–1025 (2000)
9. Asarin, E., Dang, T., Girard, A.: Reachability analysis of nonlinear systems using conservative approximation. In: Maler, O., Pnueli, A. (eds.) HSCC 2003. LNCS, vol. 2623, pp. 20–35. Springer, Heidelberg (2003). https://doi.org/10.1007/3-540-36580-X_5
10. Asarin, E., Dang, T., Girard, A.: Hybridization methods for the analysis of nonlinear systems. Acta Informatica **43**(7), 451–476 (2007)
11. Asarin, E., Dang, T., Maler, O.: The d/dt tool for verification of hybrid systems. In: Brinksma, E., Larsen, K.G. (eds.) CAV 2002. LNCS, vol. 2404, pp. 365–370. Springer, Heidelberg (2002). https://doi.org/10.1007/3-540-45657-0_30
12. Asarin, E., Maler, O., Pnueli, A.: Reachability analysis of dynamical systems having piecewise-constant derivatives. Theor. Comput. Sci. **138**, 35–66 (1995)
13. Asarin, E., Caspi, P., Maler, O.: Timed regular expressions. J. ACM **49**(2), 172–206 (2002)
14. Asarin, E., Maler, O.: On some relations between dynamical systems and transition systems. In: Abiteboul, S., Shamir, E. (eds.) ICALP 1994. LNCS, vol. 820, pp. 59–72. Springer, Heidelberg (1994). https://doi.org/10.1007/3-540-58201-0_58
15. Asarin, E., Maler, O.: Achilles and the tortoise climbing up the arithmetical hierarchy. In: Thiagarajan, P.S. (ed.) FSTTCS 1995. LNCS, vol. 1026, pp. 471–483. Springer, Heidelberg (1995). https://doi.org/10.1007/3-540-60692-0_68
16. Asarin, E., Maler, O., Pnueli, A.: Symbolic controller synthesis for discrete and timed systems. In: Antsaklis, P., Kohn, W., Nerode, A., Sastry, S. (eds.) HS 1994. LNCS, vol. 999, pp. 1–20. Springer, Heidelberg (1995). https://doi.org/10.1007/3-540-60472-3_1
17. Asarin, E., Pace, G., Schneider, G., Yovine, S.: SPeeDI—a verification tool for polygonal hybrid systems. In: Brinksma, E., Larsen, K.G. (eds.) CAV 2002. LNCS, vol. 2404, pp. 354–359. Springer, Heidelberg (2002). https://doi.org/10.1007/3-540-45657-0_28

18. Bak, S., Duggirala, P.S.: HyLaa: a tool for computing simulation-equivalent reachability for linear systems. In: Proceedings of the 20th International Conference on Hybrid Systems: Computation and Control. ACM (2017)

19. Bak, S., Tran, H.-D., Johnson, T.T.: Numerical verification of affine systems with up to a billion dimensions. In: Proceedings of the 22nd ACM International Conference on Hybrid Systems: Computation and Control, HSCC 2019, pp. 23–32. ACM (2019)

20. Bakhirkin, A., Ferrère, T., Nickovic, D., Maler, O., Asarin, E.: Online timed pattern matching using automata. In: Jansen, D.N., Prabhakar, P. (eds.) FORMATS 2018. LNCS, vol. 11022, pp. 215–232. Springer, Cham (2018). https://doi.org/10.1007/978-3-030-00151-3_13

21. Bartocci, E., et al.: Specification-based monitoring of cyber-physical systems: a survey on theory, tools and applications. In: Bartocci, E., Falcone, Y. (eds.) Lectures on Runtime Verification. LNCS, vol. 10457, pp. 135–175. Springer, Cham (2018). https://doi.org/10.1007/978-3-319-75632-5_5

22. Bogomolov, S., Forets, M., Frehse, G., Potomkin, K., Schilling, C.: JuliaReach: a toolbox for set-based reachability. In: Proceedings of the 22Nd ACM International Conference on Hybrid Systems: Computation and Control, HSCC 2019, pp. 39–44. ACM, New York (2019)

23. Bouissou, O., Mimram, S., Chapoutot, A.: HYSON: set-based simulation of hybrid systems. In: Proceedings - IEEE International Symposium on Rapid System Prototyping, RSP, pp. 79–85, October 2012

24. Bournez, O., Graça, D.S., Pouly, A.: Turing machines can be efficiently simulated by the general purpose analog computer. In: Chan, T.-H.H., Lau, L.C., Trevisan, L. (eds.) TAMC 2013. LNCS, vol. 7876, pp. 169–180. Springer, Heidelberg (2013). https://doi.org/10.1007/978-3-642-38236-9_16

25. Bournez, O., Maler, O., Pnueli, A.: Orthogonal polyhedra: representation and computation. In: Vaandrager, F.W., van Schuppen, J.H. (eds.) HSCC 1999. LNCS, vol. 1569, pp. 46–60. Springer, Heidelberg (1999). https://doi.org/10.1007/3-540-48983-5_8

26. Branicky, M.S.: Studies in hybrid systems: modelling, analysis, and control. Ph.D. thesis, Massachusetts Institute of Techology (1995)

27. Chen, X., Ábrahám, E., Sankaranarayanan, S.: Flow*: an analyzer for non-linear hybrid systems. In: Sharygina, N., Veith, H. (eds.) CAV 2013. LNCS, vol. 8044, pp. 258–263. Springer, Heidelberg (2013). https://doi.org/10.1007/978-3-642-39799-8_18

28. Chutinan, A., Krogh, B.H.: Verification of polyhedral-invariant hybrid automata using polygonal flow pipe approximations. In: Vaandrager, F.W., van Schuppen, J.H. (eds.) HSCC 1999. LNCS, vol. 1569, pp. 76–90. Springer, Heidelberg (1999). https://doi.org/10.1007/3-540-48983-5_10

29. Collins, P., Bresolin, D., Geretti, L., Villa, T.: Computing the evolution of hybrid systems using rigorous function calculus. In: Proceedings of the 4th IFAC Conference on Analysis and Design of Hybrid Systems (ADHS12), Eindhoven, The Netherlands (2012)

30. Cotton, S., Maler, O., Legriel, J., Saidi, S.: Multi-criteria optimization for mapping programs to multi-processors. In: 2011 6th IEEE International Symposium on Industrial Embedded Systems (SIES), SIES 2011, Vasteras, Sweden, 15–17 June 2011, pp. 9–17. IEEE (2011)

31. Dang, T., Le Guernic, C., Maler, O.: Computing reachable states for nonlinear biological models. Theor. Comput. Sci. **412**(21), 2095–2107 (2011)

32. Dang, T., Maler, O.: Reachability analysis via face lifting. In: Henzinger, T.A., Sastry, S. (eds.) HSCC 1998. LNCS, vol. 1386, pp. 96–109. Springer, Heidelberg (1998). https://doi.org/10.1007/3-540-64358-3_34

33. Dang, T., Maler, O., Testylier, R.: Accurate hybridization of nonlinear systems. In: Proceedings of the 13th ACM International Conference on Hybrid Systems: Computation and Control, HSCC 2010, Stockholm, Sweden, 12–15 April 2010, pp. 11–20. ACM (2010)

34. David, A., Larsen, K.G., Legay, A., Mikučionis, M., Poulsen, D.B.: Uppaal SMC tutorial. Int. J. Softw. Tools Technol. Transf. **17**(4), 397–415 (2015)

35. Donzé, A., Fanchon, E., Gattepaille, L.M., Maler, O., Tracqui, P.: Robustness analysis and behavior discrimination in enzymatic reaction networks. PLoS ONE **6**, e24246 (2011)

36. Dreossi, T.: Sapo: reachability computation and parameter synthesis of polynomial dynamical systems. In: Proceedings of the 20th International Conference on Hybrid Systems: Computation and Control, HSCC 2017, Pittsburgh, PA, USA, 18–20 April 2017, pp. 29–34 (2017)

37. Duggirala, P.S., Mitra, S., Viswanathan, M., Potok, M.: C2E2: a verification tool for stateflow models. In: Baier, C., Tinelli, C. (eds.) TACAS 2015. LNCS, vol. 9035, pp. 68–82. Springer, Heidelberg (2015). https://doi.org/10.1007/978-3-662-46681-0_5

38. Frehse, G., et al.: SpaceEx: scalable verification of hybrid systems. In: Gopalakrishnan, G., Qadeer, S. (eds.) CAV 2011. LNCS, vol. 6806, pp. 379–395. Springer, Heidelberg (2011). https://doi.org/10.1007/978-3-642-22110-1_30

39. Girard, A.: Reachability of uncertain linear systems using zonotopes. In: Morari, M., Thiele, L. (eds.) HSCC 2005. LNCS, vol. 3414, pp. 291–305. Springer, Heidelberg (2005). https://doi.org/10.1007/978-3-540-31954-2_19

40. Girard, A., Le Guernic, C.: Zonotope/hyperplane intersection for hybrid systems reachability analysis. In: Egerstedt, M., Mishra, B. (eds.) HSCC 2008. LNCS, vol. 4981, pp. 215–228. Springer, Heidelberg (2008). https://doi.org/10.1007/978-3-540-78929-1_16

41. Girard, A., Le Guernic, C., Maler, O.: Efficient computation of reachable sets of linear time-invariant systems with inputs. In: Hespanha, J.P., Tiwari, A. (eds.) HSCC 2006. LNCS, vol. 3927, pp. 257–271. Springer, Heidelberg (2006). https://doi.org/10.1007/11730637_21

42. Greenstreet, M.R., Mitchell, I.: Integrating projections. In: Henzinger, T.A., Sastry, S. (eds.) HSCC 1998. LNCS, vol. 1386, pp. 159–174. Springer, Heidelberg (1998). https://doi.org/10.1007/3-540-64358-3_38

43. Greenstreet, M.R., Mitchell, I.: Reachability analysis using polygonal projections. In: Vaandrager, F.W., van Schuppen, J.H. (eds.) HSCC 1999. LNCS, vol. 1569, pp. 103–116. Springer, Heidelberg (1999). https://doi.org/10.1007/3-540-48983-5_12

44. Greenstreet, M.R.: Verifying safety properties of differential equations. In: Alur, R., Henzinger, T.A. (eds.) CAV 1996. LNCS, vol. 1102, pp. 277–287. Springer, Heidelberg (1996). https://doi.org/10.1007/3-540-61474-5_76

45. Le Guernic, C.: Reachability analysis of hybrid systems with linear continuous dynamics. (Calcul d'Atteignabilité des Systèmes Hybrides à Partie Continue Linéaire). Ph.D. thesis, Joseph Fourier University, Grenoble, France (2009)

46. Le Guernic, C., Girard, A.: Reachability analysis of hybrid systems using support functions. In: Bouajjani, A., Maler, O. (eds.) CAV 2009. LNCS, vol. 5643, pp. 540–554. Springer, Heidelberg (2009). https://doi.org/10.1007/978-3-642-02658-4_40

47. Henzinger, T.A., Horowitz, B., Majumdar, R., Wong-Toi, H.: Beyond HYTECH: hybrid systems analysis using interval numerical methods. In: Lynch, N., Krogh, B.H. (eds.) HSCC 2000. LNCS, vol. 1790, pp. 130–144. Springer, Heidelberg (2000). https://doi.org/10.1007/3-540-46430-1_14

48. Henzinger, T.A., Kopke, P.W., Puri, A., Varaiya, P.: What's decidable about hybrid automata? J. Comput. Syst. Sci. **57**(1), 94–124 (1998)

49. Henzinger, T.A., Manna, Z., Pnueli, A.: Timed transition systems. In: de Bakker, J.W., Huizing, C., de Roever, W.P., Rozenberg, G. (eds.) REX 1991. LNCS, vol. 600, pp. 226–251. Springer, Heidelberg (1992). https://doi.org/10.1007/BFb0031995

50. Ibtissem, B.M., Norman, H., Stefan, K.: HyReach: a reachability tool for linear hybrid systems based on support functions. In: ARCH Workshop (2016)

51. Karagiannis, E.D., Popel, A.S.: A theoretical model of type I collagen proteolysis by matrix metalloproteinase (MMP) 2 and membrane type 1 MMP in the presence of tissue inhibitor of metalloproteinase 2. J. Biol. Chem. **279**(37), 39106–39114 (2004)

52. Kempf, J.-F., Bozga, M., Maler, O.: As soon as probable: optimal scheduling under stochastic uncertainty. In: Piterman, N., Smolka, S.A. (eds.) TACAS 2013. LNCS, vol. 7795, pp. 385–400. Springer, Heidelberg (2013). https://doi.org/10.1007/978-3-642-36742-7_27

53. Kong, S., Gao, S., Chen, W., Clarke, E.: dReach: δ-reachability analysis for hybrid systems. In: Baier, C., Tinelli, C. (eds.) TACAS 2015. LNCS, vol. 9035, pp. 200–205. Springer, Heidelberg (2015). https://doi.org/10.1007/978-3-662-46681-0_15

54. Kurzhanski, A., Valyi, I.: Ellipsoidal Calculus for Estimation and Control. Birkhauser, New York (1997)

55. Kurzhanskiy, A.A., Varaiya, P.: Ellipsoidal toolbox (ET). In: Proceedings of 45th IEEE Conference on Decision and Control (2006)

56. Kvasnica, M., Grieder, P., Baotić, M., Morari, M.: Multi-parametric toolbox (MPT). In: Alur, R., Pappas, G.J. (eds.) HSCC 2004. LNCS, vol. 2993, pp. 448–462. Springer, Heidelberg (2004). https://doi.org/10.1007/978-3-540-24743-2_30

57. Lafferriere, G., Pappas, G., Yovine, S.: Reachability computation for linear systems. In: Proceedings of the 14th IFAC World Congress, vol. E, pp. 7–12 (1999)

58. Maler, O., Krogh, B.H., Mahfoudh, M.: On control with bounded computational resources. In: Damm, W., Olderog, E.-R. (eds.) FTRTFT 2002. LNCS, vol. 2469, pp. 147–162. Springer, Heidelberg (2002). https://doi.org/10.1007/3-540-45739-9_11

59. Maler, O., Manna, Z., Pnueli, A.: Prom timed to hybrid systems. In: de Bakker, J.W., Huizing, C., de Roever, W.P., Rozenberg, G. (eds.) REX 1991. LNCS, vol. 600, pp. 447–484. Springer, Heidelberg (1992). https://doi.org/10.1007/BFb0032003

60. Maler, O., Nickovic, D.: Monitoring temporal properties of continuous signals. In: Lakhnech, Y., Yovine, S. (eds.) FORMATS/FTRTFT-2004. LNCS, vol. 3253, pp. 152–166. Springer, Heidelberg (2004). https://doi.org/10.1007/978-3-540-30206-3_12

61. Maler, O., Pnueli, A.: Reachability analysis of planar multi-linear systems. In: Courcoubetis, C. (ed.) CAV 1993. LNCS, vol. 697, pp. 194–209. Springer, Heidelberg (1993). https://doi.org/10.1007/3-540-56922-7_17

62. Maler, O., Pnueli, A., Sifakis, J.: On the synthesis of discrete controllers for timed systems (an extended abstract). In: Mayr, E.W., Puech, C. (eds.) STACS 1995. LNCS, vol. 900, pp. 229–242. Springer, Heidelberg (1995). https://doi.org/10.1007/3-540-59042-0_76

63. Mitchell, I., Tomlin, C.J.: Level set methods for computation in hybrid systems. In: Lynch, N., Krogh, B.H. (eds.) HSCC 2000. LNCS, vol. 1790, pp. 310–323. Springer, Heidelberg (2000). https://doi.org/10.1007/3-540-46430-1_27

64. Mitchell, I.M., Templeton, J.A.: A toolbox of Hamilton-Jacobi solvers for analysis of nondeterministic continuous and hybrid systems. In: Morari, M., Thiele, L. (eds.) HSCC 2005. LNCS, vol. 3414, pp. 480–494. Springer, Heidelberg (2005). https://doi.org/10.1007/978-3-540-31954-2_31

65. Platzer, A., Quesel, J.-D.: KeYmaera: a hybrid theorem prover for hybrid systems (system description). In: Armando, A., Baumgartner, P., Dowek, G. (eds.) IJCAR 2008. LNCS (LNAI), vol. 5195, pp. 171–178. Springer, Heidelberg (2008). https://doi.org/10.1007/978-3-540-71070-7_15

66. Pourcelot, E., et al.: Cellular iron regulation in animals: need and use of suitable models, pp. 73–89. Karlsruher Institut für Technologie (KIT), January 2014

67. Prabhakar, P., Soto, M.G.: AVERIST: an algorithmic verifier for stability. Electron. Notes Theor. Comput. Sci. **317**, 133–139 (2015). The Seventh and Eighth International Workshops on Numerical Software Verification (NSV)

68. Prabhakar, P., Viswanathan, M.: On the decidability of stability of hybrid systems. In: Proceedings of the 16th International Conference on Hybrid Systems: Computation and Control, HSCC 2013, 8–11 April 2013, pp. 53–62. ACM, Philadelphia (2013)

69. Salah, R.B., Bozga, M., Maler, O.: Compositional timing analysis. In: Proceedings of the 9th ACM & IEEE International Conference on Embedded Software, EMSOFT 2009, Grenoble, France, 12–16 October 2009, pp. 39–48. ACM (2009)

70. Stoma, S., Donzé, A., Bertaux, F., Maler, O., Batt, G.: STL-based analysis of trail-induced apoptosis challenges the notion of type I/type II cell line classification. PLoS Comput. Biol. **9**(5), e1003056 (2013)

71. Tendulkar, P., Poplavko, P., Maler, O.: Symmetry breaking for multi-criteria mapping and scheduling on multicores. In: Braberman, V., Fribourg, L. (eds.) FORMATS 2013. LNCS, vol. 8053, pp. 228–242. Springer, Heidelberg (2013). https://doi.org/10.1007/978-3-642-40229-6_16

72. Testylier, R., Dang, T.: NLTOOLBOX: a library for reachability computation of nonlinear dynamical systems. In: Van Hung, D., Ogawa, M. (eds.) ATVA 2013. LNCS, vol. 8172, pp. 469–473. Springer, Cham (2013). https://doi.org/10.1007/978-3-319-02444-8_37

73. Tomlin, C., Lygeros, J., Sastry, S.: Conflict resolution for air traffic management: a study in multi-agent hybrid systems. IEEE Trans. Autom. Control **43**(4), 509–521 (1998)

74. Varaiya, P.: Reach set computation using optimal control. In: Proceedings of KIT Workshop, pp. 377–383. Verimag, Grenoble (1998)

75. Vempati, P., Karagiannis, E.D., Popel, A.S.: A biochemical model of matrix metalloproteinase 9 activation and inhibition. J. Biol. Chem. **282**(52), 37585–37596 (2007)

Reaction Networks, Oscillatory Motifs and Parameter Estimation in Biochemical Systems

Igor Schreiber[1]([✉])[ID], František Muzika[1][ID], and Jan Červený[2][ID]

[1] University of Chemistry and Technology, Prague, Czech Republic
`igor.schreiber@vscht.cz`
[2] Global Change Research Institute, Czech Academy of Sciences,
Brno, Czech Republic

Abstract. We outline an approach to analysis of dynamics of biosystems formulated as reaction networks. In particular, we discuss stability analysis provided that stoichiometric equations are given for each reaction step together with power law rate expressions. Based on stoichiometry alone, the network at stationary state can be decomposed into elementary subnetworks (elementary modes, extreme currents, fluxes). Assuming power law kinetics, the capacity of the elementary subnetworks for displaying dynamical instabilities, such as bistability and oscillations, is evaluated. These subnetworks are then suitably combined to form the entire network satisfying certain stability constraints implied by experiments. Specifically, we assume that an experimentally measured biosystem represented by a reaction network displays an experimentally observed change from a steady state to oscillations. For the assumed reaction mechanism only a limited set kinetic parameters is known. In contrast, input/output parameters are known from the experiment. The set of unknown kinetic parameters may be estimated by finding a suitable linear combination of elementary modes via linear optimization so that the dynamics displayed by the model fits the experimentally observed behavior. Moreover, reaction network theory is useful in identifying subnetworks that are destabilizing the steady state to yield oscillations. Such subnetworks are called oscillatory motifs and possess a characteristic topology. As an example, we analyze a carbon-nitrogen metabolism of cyanobacteria and examine its oscillatory dynamics.

Keywords: Reaction networks · Oscillatory dynamics ·
Kinetic parameter estimation

1 Introduction

Stoichiometric network analysis (SNA) [3] examines stability of steady states of stoichiometric reaction networks. Model equations of such systems possess a pseudolinear form enabling the network at steady state to be decomposed into

© Springer Nature Switzerland AG 2019
M. Češka and N. Paoletti (Eds.): HSB 2019, LNBI 11705, pp. 30–41, 2019.
https://doi.org/10.1007/978-3-030-28042-0_3

elementary subnetworks (elementary fluxes, extreme currents). The elementary subnetworks can be linearly combined using arbitrarily chosen non-negative coefficients, producing the full network. Each of these coefficients represent degree of coupling of the relevant subnetwork to the network. The decomposition of a reaction network into elementary subnetworks has been widely applied to get insight into large networks, such as genome-scale metabolic networks [17]. However, in such networks kinetics is not specified and the steady state of the balanced network is tacitly assumed to be stable.

In contrast, SNA assumes that, in addition to stoichiometry, the rate equations for all reaction steps are given in terms of power law kinetics. This in turn allows to identify positive and negative feedbacks and draw conclusions about steady state instabilities of elementary subnetworks, or their linear combinations. An advantage of this analysis is that it does not require prior knowledge of rate coefficients and steady state concentrations of participating species. As a result, capacity of a given (sub)network for an instability of steady state is determined by identifying certain species and the range of their steady state concentrations, where the instability is manifested. Such instability may lead to bistability or oscillations depending on the nature of negative feedback involved. Given that an elementary subnetwork, or another small subnetwork, is unstable, its capacity for instability is inherited by the entire network provided that the unstable subnetwork plays a dominant role in the network, i.e. it is sufficiently coupled.

This kind of stability analysis proved useful when applied to chemical oscillators [4–6,19,22]. It allows for identification of a core subnetwork that gives rise to an oscillatory instability for each of the examined oscillators. The core subnetworks may share certain topological features, which allows for a categorization of chemical oscillators [5,19], each category being represented by a prototype or a motif. At the same time, species within each prototype can be classified based on the role they play in generating oscillations. Furthermore, when describing an experimental system in terms of component mass balances and kinetic rate expressions, the coupling coefficients and steady state concentrations (convex parameters) must be consistent with known rate coefficients and inflow/initial constraints (kinetic parameters).

An alternative approach toward classifying biochemical oscillators [15] does not exploit stability analysis of reaction networks; instead, it is based on qualitative identification of positive and negative feedback combined with a phase plane analysis.

In our previous work [14,18] we outlined an approach toward estimating kinetic parameters based on a suitable combination of elementary subnetworks. Here we overview the basic formulation, provide examples of distinct motifs and use the outlined approach to identify an oscillatory subnetwork in the model of carbon-nitrogen metabolism in cyanobacteria.

2   Basic Theory

A spatially homogeneous isothermal chemical oscillator is described by its stoichiometry and kinetics as follows. Let us assume a reaction network involving m reactions and a total number of species n^{tot},

$$\nu_{1j}^{L}A_1 + \cdots + \nu_{n^{tot}j}^{L}A_{n^{tot}} \longrightarrow \nu_{1j}^{R}A_1 + \cdots + \nu_{n^{tot}j}^{R}A_{n^{tot}} , j = 1, \cdots, m, \qquad (1)$$

where A_i are the reacting species and ν_{ij}^{L}, ν_{ij}^{R} are left and right stoichiometric coefficients. Any reversible reaction is treated as a pair of forward and backward steps. Dynamics of $n \leq n^{tot}$ species that are not inert products or being fixed are governed by a set of coupled mass balance equations which have the following pseudolinear form:

$$\frac{d\mathbf{x}}{dt} = \mathbf{N}\,\mathbf{v}(\mathbf{x}), \qquad (2)$$

where $\mathbf{x} = (x_1, \cdots, x_n)$ is the vector of concentrations of the interacting dynamical species, $\mathbf{N} = \{\Delta\nu_{ij}\} = \{\nu_{ij}^{R} - \nu_{ij}^{L}\}$ is the $(n \times m)$ stoichiometric matrix and $\mathbf{v} = (v_1, \cdots, v_m)$ is the non-negative vector of reaction rates (fluxes). The reaction rates are assumed to follow mass action kinetics,

$$v_j = k_j \prod_{i=1}^{n} x_i^{\kappa_{ij}} = k_j \bar{v}_j, \qquad (3)$$

where $\kappa_{ij} = \partial \ln v_j / \partial \ln x_i \geq 0$ is the reaction order of species i in reaction j and k_j is the corresponding rate coefficient, which may include fixed concentration(s) of pooled species and $\bar{v}_j$ is the reduced reaction rate. In vector notation we have $\mathbf{k} = (k_1, \cdots, k_m)$ and $\bar{\mathbf{v}}(\mathbf{x}) = (\bar{v}_1, \cdots, \bar{v}_m)$. For elementary reactions, $\kappa_{ij} = \nu_{ij}^{L}$, more generally $\kappa_{ij} \neq \nu_{ij}^{L}$ may be assumed for quasielementary steps. The kinetic matrix $\{\kappa_{ij}\}$ is denoted as $\mathbf{K}$. In flow systems, the inflows and outflows are included as pseudoreactions of zeroth and first order, respectively; the rate coefficient corresponding to an inflow term is $k_j = k_0 x_{i0}$ and that for an outflow is $k_j = k_0$, where k_0 is the flow (or dilution) rate and x_{i0} is the feed concentration of any inflowing species i.

At steady state Eq. (2) reduces to

$$\mathbf{N}\mathbf{v} = 0. \qquad (4)$$

Since the reaction rates are non-negative, the set of all $\mathbf{v}_s$ satisfying the steady state condition is a non-negative subset of the null space of $\mathbf{N}$ represented by an $(m - d)$-dimensional convex polyhedral cone delimited by faces of dimension $1, \cdots, (m - d) - 1$, where d is the rank of $\mathbf{N}$. One-dimensional faces (or edges) represent a set of minimal, irreducible, connected subnetworks called elementary subnetworks or extreme currents or elementary fluxes. There are f edges of the cone satisfying $f \geq m - d$. The edges should be properly normalized, a convenient way is to let the components of the rate vector corresponding to an edge sum up to 1. Endpoints of the normalized edges are apexes of a convex polytope of dimension $(m - d - 1)$. If $f = m - d$, the edges form a basis of the cone which is then called simplicial and the corresponding polytope is a simplex. Elementary fluxes can be obtained by algorithms of linear programming [9] or other efficient algorithms [20].

Let $\mathbf{E}_k$ denote a normalized rate vector corresponding to an elementary subnetwork. The set of all such subnetworks can be put into a matrix

$$\mathbf{E} = [\mathbf{E}_1, \cdots, \mathbf{E}_f]. \tag{5}$$

Any feasible rate vector $\mathbf{v}_s$ satisfying the steady state condition can be conveniently expressed as a non-negative linear combination of the elementary subnetworks,

$$\mathbf{v}_s = \mathbf{E}\,\boldsymbol{\alpha}, \quad \boldsymbol{\alpha} = (\alpha_1, \cdots, \alpha_f). \tag{6}$$

By substituting (3) into Eq. (6) the rate coefficients $\mathbf{k}$ are determined by choosing steady states $\mathbf{x}_s$. Thus, for a given set of convex parameters $(\boldsymbol{\alpha}, \mathbf{x}_s)$, kinetic parameters $\mathbf{k}$ are determined.

Using the convex parameters, the mass balances given by Eq. (2) are expressed as

$$\frac{d\mathbf{x}}{dt} = \mathbf{N}\,\mathrm{diag}\,(\mathbf{E}\,\boldsymbol{\alpha})\,(\mathrm{diag}\,\bar{\mathbf{v}}(\mathbf{x}_s))^{-1}\,\bar{\mathbf{v}}(\mathbf{x}). \tag{7}$$

Upon linearizing the r.h.s. at the steady state $\mathbf{x}_s$, the Jacobian matrix is

$$\mathbf{J} = \mathbf{N}\,\mathrm{diag}\,(\mathbf{E}\boldsymbol{\alpha})\,\mathbf{K}^T(\mathrm{diag}\,\mathbf{x}_s)^{-1} = -\mathbf{V}\,(\mathrm{diag}\,\mathbf{x}_s)^{-1}. \tag{8}$$

An instability of a steady state $\mathbf{x}_s$ can be determined by analyzing principal minors of the $(n \times n)$ matrix $\mathbf{V}$, which for a chosen (sub)network $\mathbf{E}\boldsymbol{\alpha}$ depends on stoichiometry and reaction orders but is independent of $\mathbf{x}_s$. If a principal minor of order ℓ involving a subset of indexes $i_1, \cdots, i_\ell$ of certain species is negative, then at least one eigenvalue of $\mathbf{J}$ is unstable, provided that the steady state concentrations of corresponding species $x_{i_1}^s, \cdots, x_{i_n}^s$ are sufficiently small [3]. It is sufficient to consider a leading negative minor with a minimal order ℓ since any higher order instability is derived from the minimal configuration.

3   Dominant Subnetworks and Oscillatory Motifs

Of primary importance are unstable subnetworks corresponding to edges. When coupled with other subnetworks through Eq. (6), the unstable subnetwork induces instability of the whole network provided that its coupling to the rest of the network is sufficiently strong. In other words, the unstable subnetwork is dominant. This concept can be generalized to unstable dominant faces. Specifically, a combination of two or more edges that are stable may still produce a negative principal minor, which typically occurs within an almost entire range of the relevant α_k's, i.e. the corresponding face is unstable in almost entire scope. We call such a face primary unstable. Thus, an increase of α_k for a primary unstable edge or the sum of α_k's for a primary unstable face will lead to oscillations via Hopf bifurcation [12] and is a clue to kinetic parameter estimation.

A negative principal minor represents positive feedback in the network and reflects the susceptibility of the subnetwork to possessing an unstable steady state provided that the corresponding steady state concentrations are sufficiently small. There are special cases when there is an oscillatory instability even in the absence of negative principal minor—so called negative feedback oscillators [15] used to describe simple gene regulatory networks—then more subtle criteria have to be applied to indicate instability [4,6]. However, such networks do not admit any other instability than the Hopf bifurcation, in particular they are missing a saddle-node bifurcation leading to bistability, and mathematically they represent a nongeneric case. Therefore the outlined features provide excellent guidelines in evaluating the potential of vast majority of reaction networks to undergo a dynamical instability.

When applying the SNA to oscillatory mechanisms of inorganic reactions, is has been found [5] that dominant subnetworks forming the core oscillator have only a few topological arrangements of their networks, which are called prototypes or motifs. All those possess an autocatalytic cycle, i.e. a cycle connecting species (denoted a type X) of which at least one has a stoichiometric overproduction. Also, part of the motif is a negative feedback loop involving a noncyclic species (denoted as type Z) and a removal of a type X species either by decomposition or via reaction with an inhibitory species (denoted as type Y).

More recently it has been found that some inorganic systems and many biochemical oscillators do not possess an autocatalytic cycle. Instead, their core oscillator possesses two type X-like species competing for a type Y-like species. In addition, there is a negative feedback loop involving type Z species, but all cycles present in the network are "ordinary" catalytic cycles (such as enzyme-complex-enzyme cycle) that do not directly support autocatalytic growth. Yet the network admits an instability leading to oscillations. Such a feature is called competitive autocatalysis. As with the inorganic chemical oscillators, a limited number of basic motifs are expected to constitute unstable dominant subnetworks of biochemical networks.

Examples of four prototypes are shown in Fig. 1. Cases (a) and (b) possess autocatalytic cycle and differ in the way type Z species is connected. In the case (a) it is provided via inflow and consumed by the cycle, whereas in the case (b) the type Z species is produced by the cycle and exerts negative feedback via production of the type Y species, which inhibits the autocatalytic growth. Cases (c) and (d) represent competitive autocatalysis. In the case (c) the type X species are on an "ordinary" catalytic cycle which is fed by an external source of X_1 allowing for autocatalytic-like accumulation if the type Y species is low. However, if Y is high, the autocatalysis is inhibited. In addition, the type Z species controls switches between the phase of accumulation and depletion of X species. Availability of Z is inflow-controlled and is analogous to the case (a). Case (d) is an example od topological arrangement where no cycle connects type X species. Instead, type Y species is self-regenerating. Negative feedback may be arranged in several ways, either inflow-limited external supply of Z as in the case (c) or with internal production of Z feeding back to X_2 as shown in the Figure

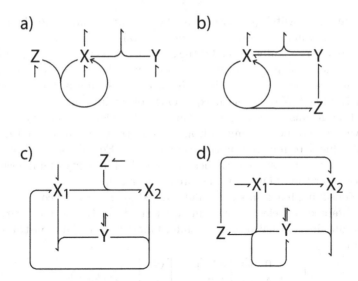

Fig. 1. Network diagrams for typical motifs of oscillatory subnetworks, (a) cyclic auto-catalysis with external source of Z, (b) cyclic autocatalysis with internal source of Z, (c) competitive autocatalysis with external source of Z, (d) competitive autocatalysis with internal source of Z.

or feeding back to X_1 (not shown). The internal production is a feature shared with the case (b). Examples of the motif in cases (a) and (b) are numerous inorganic oscillating systems [5], in particular, case (b) is the motif found in the well-known Belousov-Zhabotinsky reaction [23]. Among the enzyme systems, oxidase-peroxidase reaction [16, 21] belongs to case (a). Case (c) is a prototype of an enzyme reaction with substrate inhibition, where X_1 and X_2 are two enzyme forms, Y is the inhibitory substrate and Z is another substrate that controls the oscillations. Case (d) represents a transcriptional network with X_1 and X_2 being the activator and inhibitor, respectively, Y is the protomer and Z represents the mRNA coding for the inhibitor. Alternatively, case (d) with feedback to X_1 is found in phosphorylation cascades. Both alternatives occur simultaneously in a circadian clock model of cyanobacteria [1, 13], the phosphorylation version occurs in the model of the MAPK cascades [8, 10].

4   Parameter Estimation Using Stoichiometric Constraints

The basic idea of parameter determination is that Eq. (6) with $\mathbf{v}_s = \mathbf{v}(\mathbf{x}_s)$ can be used for finding unknown quantities including rate coefficients and steady state concentrations, given that other quantities are available, such as rate coefficients known from independent experiments or taken from literature, inflow rates, inflow concentrations and steady state concentrations obtained from an experiment at the threshold of the instability—emergence of oscillations at a

Hopf bifurcation or switch to another steady state at a saddle-node bifurcation. Depending on how many known parameters are available, only a subset of Eq. (6) can be used to express (fully or partly) the rate equations $\mathbf{v}_s = \mathbf{v}(\mathbf{x_s})$. To preserve linearity, we choose a subset of rate equations such that, upon substituting the known quantities, the rate expressions are either linear in a particular unknown or fully determined. We call such equations constraint equations. Below, the superscript fv means a fixed or given value of the relevant quantity, uv means an unknown value to be determined from the constraint equations and iv means an implied value determined from those equations in Eq. (6) that cannot be used as the constraint equations (mostly because of lack of linearity of unknowns) but can be used after the unknown values are determined.

After identifying fixed, unknown and implied quantities the constraint equations are obtained by selecting certain equations from Eq. (6) and rearranging them to obtain linear equations in a standard matrix form. The constraint equations then read

$$
\begin{bmatrix} 0 & 0 & 0 \\ \hat{\mathbf{E}}^{uv} & \mathbf{A} & 0 \\ 0 & \mathbf{B} \end{bmatrix} \begin{bmatrix} \boldsymbol{\alpha}^{uv} \\ \mathbf{k}^{uv} \\ \mathbf{x}^{uv} \end{bmatrix} = \begin{bmatrix} \mathbf{v}^{fv} \\ 0 \\ 0 \end{bmatrix} - \hat{\mathbf{E}}^{uds} \boldsymbol{\alpha}^{uds}. \tag{9}
$$

Here $\hat{\mathbf{E}}^{uv}$ is a submatrix of $\mathbf{E}$ corresponding to unknown α_k's other than the unstable dominant subnetwork, and $\hat{\mathbf{E}}^{uds}$ corresponds to the unstable dominant subnetwork. $\mathbf{A}$ and $\mathbf{B}$ are obtained from known parts of rate expressions. Further details can be found in Ref. [18].

The number of constraint equations in (9) is given by the number of reaction steps, for which we have some data in the rate expressions, in particular, rate coefficients known from experiments or literature and/or measured steady state concentrations. The number of these equations may be quite limited. On the other hand, the number of unknowns is expected to exceed the number of equations, mainly because the number of elementary subnetworks typically far exceeds the number of equations in (9) where reaction rates $\mathbf{v}^{fv}$ are known. Therefore, the system (9) is not expected to have a unique solution. Consequently, a desired solution needs to be selected by applying linear optimization using an objective function. As explained above, emergence of oscillations via Hopf bifurcation is implied by dominance of the unstable dominant subnetwork. Using Occam's razor argument, we postulate that the contributions of the elementary subnetworks other than the leading unstable subnetwork should be as small as possible at the oscillatory instability. Thus the objective function to be minimized is the sum of the α_k's of all subnetworks involved in the constraint equations other than the unstable dominant one, whose α_k^{uds} is used as a free bifurcation parameter. This parameter is stepwise varied, at each step Eq. (9) is solved until a Hopf bifurcation is found (indicated by a pair of eigenvalues of the Jacobian crossing imaginary axis). As a result, we obtain all α_k's, the unknown rate coefficients $\mathbf{k}^{uv}$ and steady state concentrations $\mathbf{x}^{uv}$ at the threshold of oscillatory instability. If there are any unknown parameters not involved in the constraint equations, they are subsequently determined as implied values from the equations in (6), which were not used as the constraint equations.

5   Case Study: Carbon-Nitrogen Metabolism in Cyanobacteria

The carbon-nitrogen (CN) metabolism of cyanobacteria [7] describes interacting pathways for the photosynthetic uptake of carbon dioxide and nitrogen fixation in cyanobacterium *Crocosphaera watsonii*. The model involves power law kinetics and describes a suspension of cells in a flow-through chemostat. A modified version of the underlying reaction network introduced in [1] is shown in Fig. 2. Notation R_k of the reactions is as in [7]. Each reaction is drawn as multiple-tail-multiple-head arrow connecting reactants and products [3]. The arrow represents both stoichiometry and power law rate expression: the label r/o at each tail of a reaction arrow represents stoichiometric coefficient/order for the corresponding reactant and the label p at each head represents the stoichiometric coefficient of the corresponding product. If there is no label on the reactant side, then $r = o = 1$, while for products p is then equal to the number of barbs at the head. Notice that stoichiometric coefficients may assume noninteger values which are adjustable parameters in some steps. Inflow/outflow streams are marked by unnumbered arrows. The two major input species N_2 and CO_2 are assumed fixed (marked by rectangles). Variables C_f and C_r correspond to functional carbon and carbohydrate compartments, N_f and N_r are functional nitrogen and nitrogen storage compartments and C_{nit} is the pool of the enzyme nitrogenase. Step R_2 corresponds to photosynthesis and step R_1 to nitrogen fixation. In the original work [7] these two processes are time-separated by controlling step R_2 by periodically varying irradiance and switching the nitrogenase activation step R_4 on/off depending on the external light to mimick day/night cycles. In a subsequent work [1] the model has been coupled with a circadian clock model to simulate interaction between the metabolism and joint action of the circadian clock and diurnal cycle. In addition, the system has been analyzed by the above outlined methods and certain modifications were proposed to avoid inconsistent

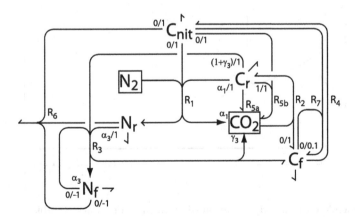

Fig. 2. Network diagram for the carbon-nitrogen metabolic model.

dynamics resulting in negative concentrations and absence of stable steady states (zero order of R_1 and R_{5b} with respect to C_r was replaced by first order and a positive order of C_f in R_2 was added).

Of interest in this work is, whether the model admits stable oscillatory regimes assuming constant light intensity and decoupled from the circadian clock. Such dynamics would correspond to the ultradian rhythm observed experimentally [2]. The analysis of the network indicates 8 elementary subnetworks, of which the subnetwork involving reactions R1, R2, R3 and R4 is unstable. For certain combinations of all the elementary subnetworks the instability generates oscillations via Hopf bifurcation. For example, at a chosen dilution rate $k_0 = D = 0.01$ (in arbitrary units) the vector $\alpha = [1, 1, 0.001, 0.1, 1, 0.01, 0.01, 0.255]$ is found to correspond to a Hopf bifurcation and both $\mathbf{x}_s$ and $\mathbf{k}$ are implied. Although this result was not obtained by employing the constraint Eq. (9) because of insufficient experimental data, it does explicitly show that oscillations are possible.

To further analyze the scope of oscillatory dynamics, we employ numerical bifurcation analysis [11]. The Hopf bifurcation point found by the network analysis can be used as a starting point for numerical determination of dependence of the steady state $\mathbf{x}_s$ on dilution rate D, see the bifurcation diagram in Fig. 3. The diagram shows the steady state value of C_f for varying D. There are three branches of steady states. The branch with zero C_f corresponds to zero ('dead') steady states, which are stable in the entire range of D. Two other branches with positive steady states coexist in the range of dilution rates from 0 to $D \approx 7$ where they merge and disappear (saddle-node bifurcation). This point corresponds to washout above which only the zero steady state exists. The lower of

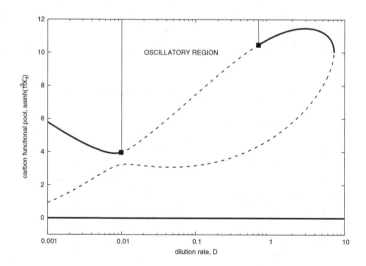

Fig. 3. Dependence of steady states on dilution rate. Full lines – stable steady state, dashed lines – unstable steady state, squares – points of the Hopf bifurcation delimiting the region of stable oscillations.

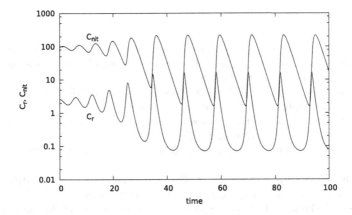

Fig. 4. Temporal periodic oscillations of C_r and C_{nit}.

the two positive branches is unstable within entire range but the upper is stable for low and high dilution rates and unstable in the middle range delimited by two points of the oscillatory instabilities (Hopf bifurcation). Within this window, stable periodic oscillations occur. Therefore the system displays bistability between positive and zero steady states for low and high D (but below the washout limit) and bistability between oscillations and zero steady state in the middle range of D. The character of oscillations is shown in Fig. 4. Periodic temporal dynamics of C_r and C_{nit} is marked by a phase shift, which spontaneously separates photosynthesis from nitrogen fixation even without intervening internal circadian or external diurnal oscillator. On a qualitative level, this provides an explanation for the ultradian cycles [2], but further modifications/extensions of the model are needed to account for quantitative description.

6   Conclusion

The approach outlined above has been successfully applied to the glucose oxidase–catalase reaction [14] and the Belousov-Zhabotinsky reaction [18]. However, target applications are expected in identifying kinetic parameters in models of biological oscillating systems including cyanobacterial rhythms as briefly mentioned above.

Acknowledgement. This work has been supported by the 18-24397S from the Czech Science Foundation.

References

1. Červený, J., Šalagovič, J., Muzika, F., Šafránek, D., Schreiber, I.: Influence of circadian clock on optimal regime of central c-n metabolism of cyanobacteria. In: Mishra, A.K., Tiwari, D.N., Rai, A.N. (eds.) Cyanobacteria: From Basic Science to Applications, chap. 9, pp. 193–206. Academic Press (2018)
2. Červený, J., Sinetová, M.A., Valledor, L., Sherman, L.A., Nedbal, L.: Ultradian metabolic rhythm in the diazotrophic cyanobacterium Cyanothece sp. ATCC 51142. PNAS **110**(32), 13210–13215 (2013)
3. Clarke, B.L.: Stability of complex reaction networks. Adv. Chem. Phys. **43**, 1–278 (1980)
4. Eiswirth, M., Bürger, J., Strasser, P., Ertl, G.: Oscillating Langmuir-Hinshelwood mechanisms. J. Phys. Chem. **100**(49), 19118–19123 (1996)
5. Eiswirth, M., Freund, A., Ross, J.: Mechanistic classification of chemical oscillators and the role of species. Adv. Chem. Phys. **80**, 127–199 (1991)
6. Errami, H., Eiswirth, M., Grigoriev, D., Seiler, W.M., Sturm, T., Weber, A.: Detection of Hopf bifurcations in chemical reaction networks using convex coordinates. J. Comput. Phys. **291**, 279–302 (2015)
7. Grimaud, G.M., Rabouille, S., Dron, A., Sciandra, A., Bernard, O.: Modelling the dynamics of carbon–nitrogen metabolism in the unicellular diazotrophic cyanobacterium crocosphaera watsonii WH8501, under variable light regimes. Ecol. Model. **291**, 121–133 (2014)
8. Hadac, O., Muzika, F., Nevoral, V., Pribyl, M., Schreiber, I.: Minimal oscillating subnetwork in the Huang-Ferrell model of the MAPK cascade. PLoS ONE **12**(6), e0178457 (2017)
9. Hadley, G.: Linear Programming. Addison-Wesley Publishing Company, Boston (1962)
10. Huang, C.Y., Ferrell, J.E.: Ultrasensitivity in the mitogen-activated protein kinase cascade. PNAS **93**(19), 10078–10083 (1996)
11. Kohout, M., Schreiber, I., Kubicek, M.: A computational tool for nonlinear dynamical and bifurcation analysis of chemical engineering problems. Comput. Chem. Eng. **26**(4–5), 517–527 (2002)
12. Marsden, J.E., McCracken, M.: The Hopf Bifurcation and Its Applications. Springer, New York (1976). https://doi.org/10.1007/978-1-4612-6374-6
13. Miyoshi, F., Nakayama, Y., Kaizu, K., Iwasaki, H., Tomita, M.: A mathematical model for the Kai-protein-based chemical oscillator and clock gene expression rhythms in cyanobacteria. J. Biol. Rhythms **22**(1), 69–80 (2007)
14. Muzika, F., Jurasek, R., Schreiberova, L., Radojkovic, V., Schreiber, I.: Identifying the oscillatory mechanism of the glucose oxidase-catalase coupled enzyme system. J. Phys. Chem. A **121**(40), 7518–7523 (2017)
15. Novak, B., Tyson, J.J.: Design principles of biochemical oscillators. Nat. Rev. Mol. Cell Biol. **9**(12), 981–991 (2008)
16. Olsen, L., Degn, H.: Chaos in an enzyme reaction. Nature **267**(5607), 177–178 (1977)
17. Palsson, B.: Systems Biology: Properties of Reconstructed Networks. Cambridge University Press, Cambridge (2006)
18. Radojkovic, V., Schreiber, I.: Constrained stoichiometric network analysis. Phys. Chem. Chem. Phys. **20**, 9910–9921 (2018)
19. Ross, J., Schreiber, I., Vlad, M.O.: Determination of Complex Reaction Mechanisms. Oxford University Press Inc., New York (2006)

20. Schilling, C.H., Letscher, D., Palsson, B.Ø.: Theory for the systemic definition of metabolic pathways and their use in interpreting metabolic function from a pathway-oriented perspective. J. Theor. Biol. **203**(3), 229–248 (2000)
21. Schreiber, I., Hung, Y.F., Ross, J.: Categorization of some oscillatory enzymatic reactions. J. Phys. Chem. **100**(20), 8556–8566 (1996)
22. Schreiber, I., Ross, J.: Mechanisms of oscillatory reactions deduced from bifurcation diagrams. J. Phys. Chem. A **107**(46), 9846–9859 (2003)
23. Zhabotinskii, A.M.: Periodic course of the oxidation of malonic acid in a solution (studies on the kinetics of Belousov's reaction). Biofizika **9**, 306–11 (1964)

Regular Papers

Fixed-Point Computation of Equilibria in Biochemical Regulatory Networks

Isabel Cristina Pérez-Verona[1], Mirco Tribastone[1], and Max Tschaikowski[2(✉)]

[1] IMT School for Advanced Studies Lucca, Lucca, Italy
{isabel.perez,mirco.tribastone}@imtlucca.it
[2] Technische Universität Wien, Vienna, Austria
max.tschaikowski@tuwien.ac.at

Abstract. The analysis of equilibria of ordinary differential equations (ODEs) that represent biochemical reaction networks is crucial in order to understand various functional properties of regulation in systems biology. In this paper, we develop a numerical algorithm to compute equilibria under the assumption that the regulatory network satisfies certain graph-theoretic conditions which lead to fixed-point iterations over an anti-monotonic function. Unlike generic approaches based on Netwon's method, our algorithm does not require the availability of the Jacobian of the ODE vector field, which may be expensive when the dimensionality of the system is large. More important, it produces an estimation (through over-approximation) of the entire set of equilibria, with the guarantee of yielding the unique equilibrium of the ODE in the case that the returned set is a singleton. We demonstrate the applicability of our algorithm to two signaling pathways of MAPK and EGFR.

1 Introduction

Ordinary differential equations (ODEs) are a fundamental dynamical model across many branches of natural and engineering sciences. In many applications, modelers are often interested in the behavior of a system when it is sufficiently away from a transient regime that depends on the conditions with which the ODEs are initialized. In this context, one is typically concerned with the study of equilibrium points; that is, given an ODE system $\dot{x} = F(x)$ over variables x, one considers the equation $0 = F(x)$, whose solution gives the states at which the ODE solution may settle over time.

In systems biology, the analysis of the equilibrium points of ODE systems that represent biochemical reaction networks carries important physical implications that are related to various regulatory phenomena; because of this, it is an area that has received considerable attention [19,27]. In this paper, we are concerned with methods to compute equilibrium points. There are different cases where this is particularly useful. For example, bifurcation analysis usually consists in producing a diagram that plots the equilibria of an ODE system as a function of a model parameter (the *bifurcation parameter*), e.g. [15]; this requires one computation for each point in the diagram.

M. Češka and N. Paoletti (Eds.): HSB 2019, LNBI 11705, pp. 45–62, 2019.
https://doi.org/10.1007/978-3-030-28042-0_4

In general, the analysis of equilibria could be approached by means of functional iteration algorithms such as Newton's method in order to find the roots of the function $F(x)$ [20]. However, there are two notable issues related to the use of such methods. The first issue is of computational nature and is due to the fact that these methods require the computation of the Jacobian of $F(x)$ at each iteration, which may be demanding when the number of ODE state variables is large. The second issue is related to the convergence properties. Indeed, Newton's method does not guarantee convergence in general. Also, it may converge to different equilibria depending on the initial guess. Finally, while it can be used to find equilibria, it cannot be used to prove that a given ODE systems has only one equilibrium point.

Contribution. The main contribution of this paper is to develop an algorithm that is designed for the computation of equilibrium points of (nonlinear) ODE systems arising from the modeling of biochemical reaction networks that aims to avoid the two aforementioned issues. We focus on models that can be represented as *regulatory networks* (RNs), similarly to [6,17,27], whereby vertices represent biomolecular species that can activate or inhibit other vertices.

Technically, the main idea behind our framework is to observe that the equilibrium equation $0 = F(x)$ mentioned above is equivalent, under certain graph-theoretic assumptions on the RN, to a fixed-point equation $x = f(x)$. A key step is the replacement of *loops* in the RN with exogenous inputs; roughly speaking, if we can uniquely characterize the equilibrium in the open-loop system through a function that enjoys some monotonicity properties, then it is possible to compute the equilibrium of the closed-loop system via a fixed-point iteration that solves $x = f(x)$. Our conditions yield that f is *anti-monotonic*, that is $f(x') \leq f(x)$ when $x \leq x'$. This implies that a composition of f with itself, i.e., $g := f \circ f$, yields a *monotonic* function. Thus, Kleene's fixed-point theorem and the fact that any fixed-point of f is necessarily a fixed-point of g, allow us to prove that all equilibria of $\dot{x} = F(x)$ are contained between the least and the greatest fixed point of g, respectively, and can be efficiently computed by fixed-point iteration, avoiding the need for the Jacobian of $F(x)$.

In practice our framework enables an efficient estimation of the set of equilibria of an ODE system underlying a RN. Apart from providing additional model insight, the estimation facilitates the use of other algorithms, e.g., it narrows down the set of initial guesses in Newton's method. More importantly, in the special case where the least and greatest fixed points coincide, the approach provides the unique equilibrium point of the ODE system. The framework has been evaluated on computational models of the well-known signaling pathways MAPK [2,18] and EGFR [5].

Related Work. Fixed-points and decomposition have been used to verify finite space dynamical systems [11]. While similar in spirit, our approach is complementary because it focuses on regulatory networks whose semantics is given in terms of differential equations. Instead, our work is closely related to [1–3,13,25,26] which also analyze RNs by suppressing feedback loops. However, [1–3,13] do not focus on the computation of the set of equilibria but prove that the ODE system converges to one of the stable equilibria. More specifically, [2,3] rule out the presence of such complex phenomena like limit cycles and ensure that convergence to unstable equilibria is only possible from a set of initial conditions with Lebesgue measure zero. Instead, [25,26] focus on the computation of the set of equilibria but do not exploit any monotonicity arguments. While this allows for a general treatment, it requires one to solve a nonlinear system of equations. Instead, under model-dependent assumptions discussed later, each iteration of the proposed fixed-point iteration algorithm can be computed in polynomial time. The closest approach to ours is [13] which proposes to decompose an in general non-monotonic system into subsystems with negative feedback. Unlike us, however, [13] does not provide an algorithm which can be used to compute (or over-approximate) the set of system's equilibria. Moreover, our algorithm is stated on the domain of regulatory networks, while [1–3,13] associate graphs to systems of differential equations [1].

2   Overview of Main Results

Before providing a formal introduction of our technique in Sect. 3, we shall present the main ideas using a running example. In particular, we consider the well-known mitogen-activated protein kinase (MAPK) signaling cascade. This is activated by several receptors, and the signal triggers the consecutive activation of several downstream protein kinases, where the last kinase can trigger cellular response such as growth, development, differentiation, proliferation, inflammation, and apoptosis. The pathway is regulated by phosphatases which dephosphorylate the kinases and interrupt the signal. This regulation occurs in intermediate steps. At the end of the pathway, a negative feedback triggers the shutdown on the activator kinase. A mutation on one of the components of the signaling pathway can cause errors in information processing interfering with the cellular response and causing diseases such as cancer, autoimmunity, diabetes, etc. Therefore, there is great interest in understanding the underlying mechanisms of this pathway, see, e.g., [4,5,16,18,21–24,28,29].

We consider a computational model which can be schematically represented as the chemical reaction network depicted in Fig. 1a, following Kholodenko [18]. It is possible to identify the three levels of the cascade (light-blue boxes), where each level consists of a cycle involving two or more interconvertible mass-preserving forms of some kinase, i.e., dephosphorylated, mono-phosphorylated, and bi-phosphorylated (the number of phosphorylated sites is indicated by the suffixes '-P' and '-PP'). The most active form regulates the downstream kinases as a catalyst for the phosphorylation reactions (e.g., the reaction with label 3), while dephosphorylation is assumed to take place spontaneously (e.g., label 2).

This model can be cast into our framework of reaction networks, where essentially we decompose the behavior into a local dynamics that describes the evolution between the different forms of a given component (i.e., a gene, a protein, a metabolite, and so on) and the network dynamics that describes the influence (that is, activation or inhibition) that one form of each such component

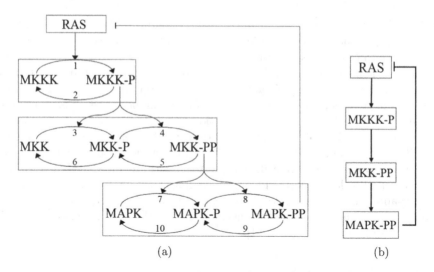

(a)   (b)

Fig. 1. Left. Original MAPK model from [18]. Right: MAPK RN. Each family of states (e.g., $\{[MKK], [MKK_P], [MKK_{PP}]\}$) is identified by the most active species (e.g., $[MKK_{PP}]$). (Color figure online)

exerts on the other components in the system. In doing so, we are essentially following a standard biochemical representation, akin to that used, among others, by Cardelli [6] as well as Tyson and Novak [27], remarking that we are not committing to specific kinetic mechanisms. Overall, this leads to a more abstract graphical representation of the RN (that will be mathematically formalized later) as depicted in Fig. 1b, where we replace each box of Fig. 1a with a single vertex (labeled with the most active form of the component) and connect the vertices through activation or inhibition arcs. The vertex label indicates the *observable*, that is, the ODE variable related to the specific component form (for instance, the most active form) that appears in the ODEs of the other components.

The main contribution of the present paper is a graph-theoretic analysis of the RN for the computation of the equilibrium points of its associated ODE system. To see this with our example, let us make the assumption of mass-action dynamics. Then, the ODE equations associated with the vertices labeled RAS and MKKK-P are thus:

$$[\dot{RAS}](t) = \lambda - \beta_0[MAPK_{PP}](t)[RAS](t) - \gamma[RAS](t) \tag{1}$$

$$[\dot{MKKK}](t) = \beta_2[MKKK_P](t) - \alpha_1[MKKK](t)[RAS](t) \tag{2}$$

$$[\dot{MKKK_P}](t) = \alpha_1[MKKK](t)[RAS](t) - \beta_2[MKKK_P](t) \tag{3}$$

where, following standard notation, species (i.e., state variable) names in square brackets refer to concentrations. In these equations, α_i, λ, β_i, and γ are positive parameters, respectively (in particular, we assume that RAS is injected into the system at a constant rate).

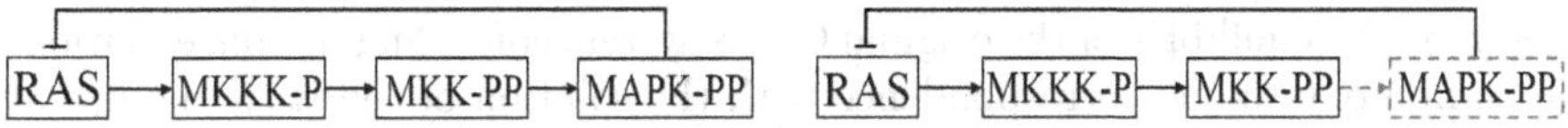

(a) Closed loop MAPK model.   (b) Open loop MAPK model.

Fig. 2. The feedback from MKK$_{PP}$ to MAPK$_{PP}$ is suppressed and vertex MAPK$_{PP}$ is interpreted as an input vertex.

The key idea of our method builds upon [2,13,26] and consists in transforming a *closed-loop* RN into an *open-loop* one by suppressing one or several edges of the original network. Vertices whose incoming edges are suppressed are treated as *input* vertices. Figure 2 visualizes this process in the case of our running example.

Under certain assumptions, the dynamical system underlying the open loop network has a unique equilibrium and can be efficiently computed *whenever a constant input is provided*. For instance, by removing the feedback loop from MAPK$_{PP}$ to RAS, we treat [MAPK$_{PP}$] as a constant input in the ODE (1) of RAS. With this, we can compute the equilibrium of RAS by setting its derivative to zero as

$$[\text{RAS}] = \frac{\lambda}{\beta_{11}[\text{MAPK}_{PP}] + \gamma} =: \phi_{\text{RAS}}([\text{MAPK}_{PP}]) \tag{4}$$

where the ϕ_{RAS} function maps the input [MAPK$_{PP}$] to the unique equilibrium point of [RAS]. We can continue this reasoning along the pathway, by interpreting [RAS] as a constant input in the ODEs (2)–(3) of MKKK and MKKK$_P$. This leads to the equilibrium

$$[\text{MKKK}_P] = \frac{\alpha_1 c_{\text{MKKK}_P}}{\alpha_1[\text{RAS}] + \beta_2} =: \phi_{\text{MKKK}_P}([\text{RAS}]), \tag{5}$$

where we have set

$$c_{\text{MKKK}_P} := [\text{MKKK}](0) + [\text{MKKK}_P](0)$$

thanks to the mass conservation property $[\text{MKKK}](t) + [\text{MKKK}_P](t) = 0$. By continuing in a similar fashion, one can derive the equilibrium formula $\phi_{\text{MKK}_{PP}}([\text{MKKK}_P])$.

While the above discussion implies that the equilibrium point of the open-loop network given in Fig. 2b can be computed efficiently, its relation to the equilibria of the original closed-loop network in Fig. 2a is not obvious. We can address this problem by *closing* the open loop network by reactivating the previously suppressed edge. Formally, this corresponds to the feedback condition [MAPK$_{PP}$] = $\phi_{\text{MAPK}_{PP}}([\text{MKK}_{PP}])$, where $\phi_{\text{MAPK}_{PP}}([\text{MKK}_{PP}])$ denotes the equilibrium value of MAPK$_{PP}$ in the case of a given constant input [MKK$_{PP}$]. Overall, we obtain the fixed-point equation

$$[\text{MAPK}_{PP}] = \phi_{\text{MAPK}_{PP}}\Big(\phi_{\text{MKK}_{PP}}\Big(\phi_{\text{MKKK}_P}\big(\phi_{\text{RAS}}([\text{MAPK}_{PP}])\big)\Big)\Big)$$

$$=: f_{\text{MAPK}_{PP}}([\text{MAPK}_{PP}]) \tag{6}$$

Therefore, the equilibria of the original ODEs system consisting of *nine* equations stands in a one-to-one correspondence with the solution of the *single* nonlinear Eq. (6). More formally, if [MAPK$_{PP}$] is the value of the MAPK$_{PP}$-coordinate of an equilibrium of the ODE system, then [MAPK$_{PP}$] solves (6). Conversely, every solution of (6) induces the equilibria [RAS], [MKKK$_P$], [MKK$_{PP}$] and [MAPK$_{PP}$] for the ODEs of RAS, MKKK$_P$, MKK$_{PP}$ and MAPK$_{PP}$, respectively, through the functions ϕ_{RAS}, ϕ_{MKKK_P}, and $\phi_{MKK_{PP}}$. Together with the equilibrium conditions imposed by the ODEs and the conservation of mass, this determines the equilibria of the remaining five ODEs.

In [25,26] the above discussion is generalized and the equilibria of a nonlinear ODE system are expressed as fixed points of a nonlinear vector function of smaller size. Unfortunately, the computation of the solution set of a system of nonlinear equations is computationally prohibitive and does not scale [25]. Instead, [13] identifies further conditions under which the dynamical system converges to the set of fixed points but does not address their automatic computation.

The present work addresses this problem by showing that the system of nonlinear equations can be often solved via a fixed-point iteration algorithm. More specifically, we identify graph-theoretic conditions under which an RN induces a fixed-point equation $x = f(x)$ such that f is *anti-monotonic*, i.e., $x \leq x'$ implies $f(x') \leq f(x)$, where $\leq$ is to be interpreted componentwise if x and $f(x)$ are vectors. For instance, let us denote by $c_{MAPK_{PP}}$ the maximal concentration attainable by MAPK in any of its forms, i.e.,

$$c_{MAPK_{PP}} := [MAPK](0) + [MAPK_P](0) + [MAPK_{PP}](0),$$

Then, for any $0 \leq x \leq x' \leq c_{MAPK_{PP}}$, where it holds that $f_{MAPK_{PP}}(x') \leq f_{MAPK_{PP}}(x)$. In the case of the MAPK model, this should not surprise because it essentially states that an increase in inhibition leads to a decrease in activation.

While anti-monotonicity does not imply in general that a fixed-point iteration $x, f(x), f(f(x)), \dots$ converges to a fixed-point of f, it ensures that the composition $g := f \circ f$ is *monotonic*, i.e., $0 \leq x \leq x' \leq c$ implies $g(x) \leq g(x')$, with c being the vector of all maximal attainable species concentrations. This and Kleenes's fixed-point theorem ensure that the sequences $0, g(0), g(g(0)), \dots$ and $c, g(c), g(g(c)), \dots$ converge to the least and the greatest fixed-point of g, respectively. Noting that any fixed-point of f has to be necessarily a fixed-point of $g = f \circ f$, this can be used to prove the following two statements.

1. Let $x^{\perp}$ and $x^{\top}$ be the least and greatest fixed-point of g, respectively. Then, any equilibrium of the ODE system underlying the RN is contained in $[x^{\perp}; x^{\top}]$.
2. If $x^* = x^{\perp} = x^{\top}$, then x^* is the unique equilibrium of the ODE system underlying the RN.

The first statement allows one to efficiently estimate the set of equilibria of the ODE system, while the second statement allows one to decide whether the ODE system has a unique equilibrium and, in the case it does, allows for an efficient computation of it.

If applied to the MAPK model, the above discussion ensures that we obtain the unique solution of the nonlinear Eq. (6) when the sequences

$$0, \qquad g_{\mathrm{MAPK_{PP}}}(0), \qquad g_{\mathrm{MAPK_{PP}}}\big(g_{\mathrm{MAPK_{PP}}}(0)\big), \qquad \cdots$$

and

$$c_{\mathrm{MAPK_{PP}}}, \qquad g_{\mathrm{MAPK_{PP}}}(c_{\mathrm{MAPK_{PP}}}), \qquad g_{\mathrm{MAPK_{PP}}}\big(g_{\mathrm{MAPK_{PP}}}(c_{\mathrm{MAPK_{PP}}})\big), \qquad \cdots$$

converge to the same value, with $g_{\mathrm{MAPK_{PP}}} := f_{\mathrm{MAPK_{PP}}} \circ f_{\mathrm{MAPK_{PP}}}$.

3   Computation of Equilibria in RNs

Notation. We denote vectors with index set I by $\mathbb{R}^I$. For $x \in \mathbb{R}^I$ and $I_0 \subseteq I$, the restriction of x to the index set I_0 is denoted by $x_{|I_0}$.

Let us now formalize the ideas and results presented in the previous section. We start by fixing the notation for a RN.

Definition 1 (Regulatory Network). *A Regulatory Network (RN) is a directed graph (V, E) where V is the set of vertices and E is the set of labeled edges, i.e., $E \subseteq V \times \{+, -\} \times V$.*

- *The set of all outgoing neighbors of vertex $i \in V$ is denoted by $\boldsymbol{out}(i)$, that is $\boldsymbol{out}(i) = \{j \in V \mid (i, \cdot, j) \in E\}$. Similarly, $\boldsymbol{in}(i)$ denotes all incoming neighbors of i.*
- *A vertex is called* activator *(resp.,* inhibitor*) if all its outgoing edges are activating (resp., inhibiting). We let V^+ and V^- be the sets of activator and inhibitor vertices, respectively.*
- *The set of inhibitors with incoming edges defines the set of core observables $\mathcal{I}$, that is, $\mathcal{I} = \{i \in V^- \mid \boldsymbol{in}(i) \neq \emptyset\}$.*

The symbols in the label set of the edges denotes activation and inhibition in the obvious way. The set of core observables $\mathcal{I}$ corresponds essentially to those vertices whose incoming edges are all suppressed and which act as exogenous inputs in the open network. Throughout the remainder of this paper we exclude the case of vertices with no outgoing edges. For a more compact notation, the set of vertices is assumed to be $V = \{1, 2, \ldots, n\}$ for some $n \geq 1$.

We now define the semantics of an RN in terms of a system of coupled ODEs. In particular, with a given vertex i we associate a set of m_i ODEs over variables $x_i^1, \ldots, x_i^{m_i}$. Essentially, each of the m_i variables $x_i^1, \ldots, x_i^{m_i}$ related to a vertex i may represent the different forms that a component can exhibit, e.g., MKKK and MKKK$_P$ in Fig. 1a. One such variable (i.e., the first component x_i^1 without loss of generality) is chosen to represent the *observable*—the only variable that may appear in the set of ODE related to the other vertices; this formalizes the idea of the labels used in the pictorial representation of the RN of, for example, Fig. 1b.

Definition 2 (RN Semantics). *Given an RN, its underlying ODE is given by associating each vertex $i \in V$ with a system of $m_i \geq 1$ ODEs over variables $x_i^1, \ldots, x_i^{m_i}$. The ODE system associated with vertex i is given by*

$$\dot{\boldsymbol{x}}_i = F_i(x_1^1, \ldots, x_{i-1}^1, \boldsymbol{x}_i, x_{i+1}^1, \ldots, x_n^1), \quad \text{with} \quad \boldsymbol{x}_j := (x_j^1, \ldots, x_j^{m_j}) \quad \text{for } 1 \leq j \leq n.$$

The ODE system underlying the RN is given by

$$\dot{\boldsymbol{x}} = F(\boldsymbol{x}), \quad \text{where} \quad \boldsymbol{x} = (\boldsymbol{x}_1, \ldots, \boldsymbol{x}_n) \quad \text{and} \quad F = (F_1, \ldots, F_n).$$

The initial condition of the RN is denoted by $\boldsymbol{x}(0)$ and assumed to be non-negative. Moreover, we assume that F is Lipschitz continuous and that the solution of $\dot{\boldsymbol{x}} = F(\boldsymbol{x})$ remains non-negative if initialized with a non-negative initial condition.

In the remainder, we avoid the use of the superscript for the observable of vertex i, i.e., we write x_i to indicate x_i^1.

We next introduce well-posed RNs. Intuitively, this identifies a local property of the network whereby the ODE system associated with each vertex enjoys a unique equilibrium when the observables of the other vertices that act as inhibitors or activators are treated as exogenous constant inputs.

Definition 3 (Well-Posed RN). *For fixed initial condition $x(0)$ and model parameters, an RN (V, E) is called* well-posed *when the following conditions are satisfied.*

(1) *V^+ and V^- form a partition of V, i.e., there are no vertices that are both activators and inhibitors.*

(2) *For all $i \in V$, the maximal value of x_i attainable across all non-negative initial conditions $\hat{\boldsymbol{x}}(0)$ satisfying $\|\hat{\boldsymbol{x}}_j(0)\|_1 = \|\boldsymbol{x}_j(0)\|_1$, where $j \in V$ is arbitrary and $\|\cdot\|_1$ denotes the L^1 norm, exists and is denoted by c_i.*

(3) *For every vertex $i \in V$, define*
   - *$\mathcal{U}_i^+ := \prod_{j \in V^+ \cap in(i)}[0; c_j]$ the set of admissible activation inputs; and*
   - *$\mathcal{U}_i^- := \prod_{j \in V^- \cap in(i)}[0; c_j]$ the set of admissible inhibition inputs.*
   Then, for every $i \in V$, $u^+ \in \mathcal{U}_i^+$ and $u^- \in \mathcal{U}_i^-$, it must hold that the equilibrium equation

$$0 = F_i(x_1, \ldots, x_{i-1}, \boldsymbol{x}_i, x_{i+1}, \ldots, x_n)$$

admits a non-negative solution that satisfies $x_j = u_j^+$ for all $j \in V^+ \cap in(i)$ and $x_k = u_k^-$ for all $k \in V^- \cap in(i)$.
Moreover, the equilibrium equation must characterize the value $\boldsymbol{x}_i$. That is, given two non-negative solutions

$$0 = F_i(x_1, \ldots, x_{i-1}, \boldsymbol{x}_i, x_{i+1}, \ldots, x_n)$$
$$0 = F_i(\hat{x}_1, \ldots, \hat{x}_{i-1}, \hat{\boldsymbol{x}}_i, \hat{x}_{i+1}, \ldots, \hat{x}_n)$$

satisfying $x_j = \hat{x}_j = u_j^+$ for all $j \in V^+ \cap in(i)$ and $x_k = \hat{x}_k = u_k^-$ for all $k \in V^- \cap in(i)$, then $\boldsymbol{x}_i = \hat{\boldsymbol{x}}_i$.
The uniqueness property ensures the well-definedness of $\phi_i(u^+, u^-) := \boldsymbol{x}_i$.

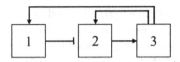

Fig. 3. An RN that is not anti-monotonic because is violates condition (*iii*).

Essentially, the equilibrium of vertex i is uniquely determined in the case one is given constant activation and inhibition inputs u^+ and u^-, respectively. Since inputs are described by the values of the observables $x = (x_1, \ldots, x_n)$, one can use x instead of the redundant (at least as far the steady-state regime is concerned) vector $\boldsymbol{x}$.

For instance, in the case of the MAPK model from Fig. 1a, Eqs. (4) and (5) show that the vertices associated with RAF and MKKP in Fig. 1b do satisfy the requirement above.

Armed with the notion of well-posedness, we are ready to introduce anti-monotonic RNs, the core concept of the present paper. The graph-theoretic conditions describing an anti-monotonic network ensure that the corresponding nonlinear fixed-point equation characterizing the equilibria, $y = f(y)$, can be constructed and is anti-monotonic, that is, it satisfies $f(y') \leq f(y)$ for all $y \leq y'$.

Definition 4 (Anti-monotonic RN). *A well-posed network (V, E) is called anti-monotonic if the following properties are satisfied.*

(*i*) *There are no inhibitors that are inhibited.*
(*ii*) *All vertices with incoming neighbors can be reached from the set of core observables $\mathcal{I}$.*
(*iii*) *The graph which arises from (V, E) by removing all inhibitors is acyclic, that is, $(V^+, E \setminus (V^- \times \{-, +\} \times V \cup V \times \{-, +\} \times V^-))$ is acyclic.*
(*iv*) *With $\phi_i(u^+, u^-)$ being as in Definition 3, ϕ_i is a continuous function that is monotonic and anti-monotonic in u^+ and u^-, respectively. That is:*
  – ϕ_i *is continuous as a function of $(u^+, u^-) \in \mathcal{U}_i^+ \times \mathcal{U}_i^-$;*
  – $\phi_i(u^-, u^+) \leq \phi_i(u^-, \hat{u}^+)$ *if $u^- \in \mathcal{U}_i^-$ and $u^+, \hat{u}^+ \in \mathcal{U}_i^+$ with $u^+ \leq \hat{u}^+$;*
  – $\phi_i(u^-, u^+) \geq \phi_i(\hat{u}^-, u^+)$ *if $u^-, \hat{u}^- \in \mathcal{U}_i^-$ and $u^+ \in \mathcal{U}_i^+$ with $u^- \leq \hat{u}^-$.*

We first remark that (*ii*) requires that all $i \notin \mathcal{I}$ depend on the core observables $\mathcal{I}$. Intuitively, this is needed to ensure that the equilibrium of the opened network can be computed from the values assigned to $\mathcal{I}$. Condition (*iii*), instead, is needed to exclude dependency deadlocks. For instance, the network depicted in Fig. 3 cannot be handled because the equilibrium of 2 depends on the equilibrium of 3 and vice versa. Overall, (*ii*) and (*iii*) ensure that f in the fixed-point equation $y = f(y)$ is well-defined, while condition (*i*) and (*iv*) imply that f is anti-monotonic.

The graph-theoretic conditions (i)–(iii) boil down to verifying that a subgraph of (V, E) has no loops, a well-known problem that can be solved in polynomial time. Instead, condition (iv) depends on the actual ODE system underlying the RN. For instance, recall that in the case of the MAPK network from Fig. 1b, we have $\phi_{\text{MKKK}_\text{P}}([\text{RAS}]) = \alpha_1 c_{\text{MKKK}_\text{P}}/(\alpha_1[\text{RAS}] + \beta_2)$. Hence, (5) fulfills condition (iv). More in general, the following can be proven.

Lemma 1. *The MAPK network from Fig. 1b satisfies (i)–(iv).*

Proof. Since ϕ_{RAS} and $\phi_{\text{MKKK}_\text{P}}$ have been covered in Sect. 2, we are left to derive the expressions $\phi_{\text{MKK}_\text{PP}}$ and $\phi_{\text{MAPK}_\text{PP}}$. Noting that $\phi_{\text{MAPK}_\text{PP}}$ is an instance of $\phi_{\text{MKK}_\text{PP}}$ with modified activation and inhibition parameters, it suffices to focus on $\phi_{\text{MKK}_\text{PP}}$ only. Standard algebraic manipulations reveal that

$$\phi_{\text{MKK}_\text{PP}}([\text{MKKK}_\text{P}]) = c_{\text{MKK}_\text{PP}} \frac{\alpha_3\alpha_4[\text{MKKK}_\text{P}]^2}{\alpha_3\alpha_4[\text{MKKK}_\text{P}]^2 + \alpha_3\beta_4[\text{MKKK}_\text{P}] + \beta_5\beta_6}$$

For completeness, we remark here that

$$\phi_{\text{MKK}}([\text{MKKK}_\text{P}]) = c_{\text{MKK}_\text{PP}} \frac{\beta_5\beta_6}{\alpha_3\alpha_4[\text{MKKK}_\text{P}]^2 + \alpha_3\beta_4[\text{MKKK}_\text{P}] + \beta_5\beta_6}$$

$$\phi_{\text{MKK}_\text{P}}([\text{MKKK}_\text{P}]) = c_{\text{MKK}_\text{PP}} \frac{\alpha_3\beta_4[\text{MKKK}_\text{P}]}{\alpha_3\alpha_4[\text{MKKK}_\text{P}]^2 + \alpha_3\beta_4[\text{MKKK}_\text{P}] + \beta_5\beta_6}$$

Since α_i and β_i are positive, it suffices to show that the function $h(y) := a_1y^2/(a_1y^2 + a_2y + a_3)$ is monotonic in y when $a_1, a_2, a_3 > 0$. A differentiation of h with respect to y yields

$$(\partial_y h)(y) = \frac{2a_1y}{a_1y^2 + a_2y + a_3} - \frac{a_1y^2(a_2 + 2a_1y)}{(a_1y^2 + a_2y + a_3)^2}$$

Algebraic manipulations reveal that $\partial_y h$ has the roots $y = 0$ and $y = -\frac{2a_3}{a_2} < 0$. Since h is non-negative and $h(0) = 0$, we infer the claim.   □

In our framework, we suppress the incoming edges of the core observables $\mathcal{I}$ and show that the equilibrium of the so-obtained open-loop network can be computed from the values assigned to $\mathcal{I}$. To do so, we will work with three index sets. First, there is the full vector $\boldsymbol{x}$. The sub-vector x of $\boldsymbol{x}$ which lives in $\mathbb{R}^V$ and tracks the observables. Finally, $x_{|\mathcal{I}}$ provides only the values of those vertices who become dangling in the open network. Our theorems work on this coarsest set $\mathcal{I}$ and show that the knowledge of those values characterizes the equilibria $\boldsymbol{x}$ of the ODE system $\dot{\boldsymbol{x}} = F(\boldsymbol{x})$ in full.

In particular, the next result ensures that Algorithm 1 computes, for a given vector $y \in \prod_{i\in\mathcal{I}}[0; c_i]$, a value $f(y) \in \prod_{i\in V}[0; c_i]$ such that the fixed-points of $y \mapsto f_{|\mathcal{I}}(y)$ stand in an one-to-one correspondence with the equilibria of the ODE system $\dot{\boldsymbol{x}} = F(\boldsymbol{x})$ of the RN. Thus, if $y = f_{|\mathcal{I}}(y)$, then the unique equilibrium $\boldsymbol{x}$ associated to y can be computed from y.

For instance, in the case of the MAPK network from Fig. 1b, $\mathcal{I} = \{\text{MAPK}_{\text{PP}}\}$ and $y \mapsto f_{|\mathcal{I}}(y)$ rewrites to $[\text{MAPK}_{\text{PP}}] \mapsto f_{\text{MAPK}_{\text{PP}}}([\text{MAPK}_{\text{PP}}])$, thus resembling (6). The function value $f([\text{MAPK}_{\text{PP}}])$, instead, has the four components

$$f_{\text{RAS}}([\text{MAPK}_{\text{PP}}]) = \phi_{\text{RAS}}([\text{MAPK}_{\text{PP}}])$$
$$f_{\text{MKKK}_{\text{P}}}([\text{MAPK}_{\text{PP}}]) = \phi_{\text{MKKK}_{\text{P}}}(f_{\text{RAS}}([\text{MAPK}_{\text{PP}}]))$$
$$f_{\text{MKK}_{\text{PP}}}([\text{MAPK}_{\text{PP}}]) = \phi_{\text{MKK}_{\text{PP}}}(f_{\text{MKKK}_{\text{P}}}([\text{MAPK}_{\text{PP}}]))$$
$$f_{\text{MAPK}_{\text{PP}}}([\text{MAPK}_{\text{PP}}]) = \phi_{\text{MAPK}_{\text{PP}}}(f_{\text{MKK}_{\text{PP}}}([\text{MAPK}_{\text{PP}}])),$$

where ϕ_{RAS} and $\phi_{\text{MKKK}_{\text{P}}}$ are given in (4) and (5), respectively (while $\phi_{\text{MKK}_{\text{PP}}}$ and $\phi_{\text{MAPK}_{\text{PP}}}$ are derived in the proof of Lemma 1). If $[\text{MAPK}_{\text{PP}}]$ satisfies $[\text{MAPK}_{\text{PP}}] = f_{\text{MAPK}_{\text{PP}}}([\text{MAPK}_{\text{PP}}])$, we set

$$[\text{RAS}] := f_{\text{RAS}}([\text{MAPK}_{\text{PP}}]) \qquad [\text{MKKK}_{\text{P}}] := f_{\text{MKKK}_{\text{P}}}([\text{MAPK}_{\text{PP}}])$$
$$[\text{MKK}_{\text{PP}}] := f_{\text{MKK}_{\text{PP}}}([\text{MAPK}_{\text{PP}}]) \qquad [\text{MAPK}_{\text{PP}}] := f_{\text{MAPK}_{\text{PP}}}([\text{MAPK}_{\text{PP}}])$$

With this, the remaining components of the equilibrium, $[\text{MKKK}]$, $[\text{MKK}]$, $[\text{MKK}_{\text{P}}]$, $[\text{MAPK}]$ and $[\text{MAPK}_{\text{P}}]$, are determined by the corresponding equilibrium equations. In the case of $[\text{MKKK}]$, $[\text{MKKK}_{\text{P}}]$ and $[\text{MKKK}_{\text{PP}}]$, for instance, it holds that (see proof of Lemma 1)

$$[\text{MKK}] = c_{\text{MKK}_{\text{PP}}} \frac{\beta_5\beta_6}{\alpha_3\alpha_4[\text{MKKK}_{\text{P}}]^2 + \alpha_3\beta_4[\text{MKKK}_{\text{P}}] + \beta_5\beta_6}$$

$$[\text{MKK}_{\text{P}}] = c_{\text{MKK}_{\text{PP}}} \frac{\alpha_3\beta_4[\text{MKKK}_{\text{P}}]}{\alpha_3\alpha_4[\text{MKKK}_{\text{P}}]^2 + \alpha_3\beta_4[\text{MKKK}_{\text{P}}] + \beta_5\beta_6}$$

$$[\text{MKK}_{\text{PP}}] = c_{\text{MKK}_{\text{PP}}} \frac{\alpha_3\alpha_4[\text{MKKK}_{\text{P}}]^2}{\alpha_3\alpha_4[\text{MKKK}_{\text{P}}]^2 + \alpha_3\beta_4[\text{MKKK}_{\text{P}}] + \beta_5\beta_6},$$

where $c_{\text{MKK}_{\text{PP}}} = [\text{MKK}](0) + [\text{MKK}_{\text{P}}](0) + [\text{MKK}_{\text{PP}}](0)$. Noting that the above values $[\text{MKK}]$, $[\text{MKK}_{\text{P}}]$ and $[\text{MKK}_{\text{PP}}]$ depend only on $[\text{MKKK}_{\text{P}}]$, we observe that the equilibrium of vertex MKK_{P} is fully determined by $[\text{MKKK}_{\text{P}}]$. Hence, vertex $[\text{MKKK}_{\text{P}}]$ satisfies condition (3) of Definition 3 which requires that the equilibrium equation of vertex $i \in V$,

$$0 = F_i(x_1, \ldots, x_{i-1}, \boldsymbol{x}_i, x_{i+1}, \ldots, x_n),$$

characterizes the value of $\boldsymbol{x}_i$ (with respect to the given inputs).

Remark 1. Condition *(iv)* of Definition 4 is usually established by deriving closed-form expressions for $(\phi_i)_{i \in V}$. Note that this problem is model-dependent. In the case of the MAPK model, for instance, it is sufficient to provide three (instead of four) closed-form expressions because $\phi_{\text{MAPK}_{\text{PP}}}$ is an instance of $\phi_{\text{MKK}_{\text{PP}}}$ with modified activation and inhibition parameters. In the model studied in Sect. 4, instead, it suffices to derive only one closed-form expression because all vertices share a common ODE semantics.

Theorem 1. *Assume that (V, E) is an anti-monotonic RN. Then, Algorithm 1 computes, for any vector $y \in \prod_{i \in \mathcal{I}}[0; c_i]$, a value $f(y) \in \prod_{i \in V}[0; c_i]$ such that:*

(1) f is anti-monotonic, i.e., $0 \le y \le y' \le c_{|\mathcal{I}}$ implies $0 \le f(y') \le f(y) \le c$.
(2) There is a one-to-one correspondence between the equilibria of F and the fixed-points of $f_{|\mathcal{I}}$. More formally
  - If $\boldsymbol{x}$ is such that $0 = F(\boldsymbol{x})$, then $f(\boldsymbol{x}_{|\mathcal{I}}) = \boldsymbol{x}_{|\mathcal{I}}$.
  - If $y \in \prod_{i \in \mathcal{I}}[0; c_i]$ satisfies $f_{|\mathcal{I}}(y) = y$, then there exists a unique $\boldsymbol{x}$ such that $0 = F(\boldsymbol{x})$ and $y = \boldsymbol{x}_{|\mathcal{I}}$. Moreover, $f(y) = \boldsymbol{x}_{|V}$.
(3) If each ϕ_i can be computed in polynomial time, then the same applies to f.

Statement (2) ensures that it suffices to find all fixed-points of $f_{|\mathcal{I}}$ in order to determine the equilibria of $\dot{\boldsymbol{x}} = F(\boldsymbol{x})$. In particular, given a fixed-point y of $f_{|\mathcal{I}}$, the vector $f(y)$ provides the observables V with values, i.e., $x_i = f_i(y)$. Hence, $\boldsymbol{x}_i$ can be obtained by solving the equilibrium equations, see discussion preceding Theorem 1. In the case of the MAPK model, statement (3) ensures that each iteration of the algorithm can be computed in polynomial time.

In Algorithm 1, any vector $y \in \prod_{i \in \mathcal{I}}[0; c_i]$ is processed in three stages. In the first stage, the for loop from line 2, the algorithm sets the f value of each vertex in $\mathcal{I}$ (i.e., $f_i(y) := y_i$ for all $i \in \mathcal{I}$) and computes the f values of all vertices that have no incoming neighbors (note that the equilibrium function of any vertex $i \in V$ with $\text{in}(i) = \emptyset$ has no inputs, hence ϕ_i is merely a constant).

In the second stage, the algorithm computes the f values of all remaining vertices, that is vertices that are activators with at least one incoming neighbor. The underlying computation is carried out using the while loop from line 10. This is because the f values in question have to be computed in a specific order that is dictated by the graph. For instance, in the case of the MAPK network from Fig. 1b, $f_{\text{RAS}}([\text{MAPK}_{\text{PP}}])$ has to be computed before $f_{\text{MKKK}_{\text{P}}}([\text{MAPK}_{\text{PP}}])$ can be computed. Because of this, $f_{\text{MKKK}_{\text{P}}}([\text{MAPK}_{\text{PP}}])$ is computed during the second iteration, while $f_{\text{RAS}}([\text{MAPK}_{\text{PP}}])$ is obtained in the first iteration.

The third and final stage of the algorithm is given by the for loop in line 20. There, the algorithm computes the f values of all $i \in \mathcal{I}$ using the previously computed f values. This intuitively corresponds to a closing of the opened network, see Sect. 2. In particular, if the f values computed during the for loop in line 20 coincide with $y \in \prod_{i \in \mathcal{I}}[0; c_i]$, then $y = f_{|\mathcal{I}}(y)$.

If applied to the MAPK network from Fig. 1b, Algorithm 1 repeats the computational steps from Sect. 2.

Armed with Theorem 1, we are in a position to state our main result.

Theorem 2. *Assume that (V, E) is an anti-monotonic RN and let $\mathcal{I}$ and f be as in Theorem 1. Define $g(y) := f_{|\mathcal{I}}(f_{|\mathcal{I}}(y))$ for any $y \in \prod_{i \in \mathcal{I}}[0; c_i]$. Then, the following holds true.*

(1) *Function $f_{|\mathcal{I}} : \prod_{i \in \mathcal{I}}[0; c_i] \to \prod_{i \in \mathcal{I}}[0; c_i]$ has at least one fixed-point.*
(2) *Function g is monotonic, i.e., $0 \le y \le y' \le c_{|\mathcal{I}}$ implies $0 \le g(y) \le g(y') \le c$.*
(3) *Sequence $0_{|\mathcal{I}}, g(0_{|\mathcal{I}}), g(g(0_{|\mathcal{I}})), \ldots$ converges to $x^{\perp}$, the least fixed-point of g.*
(4) *Sequence $c_{|\mathcal{I}}, g(c_{|\mathcal{I}}), g(g(c_{|\mathcal{I}})), \ldots$ tends to $x^{\top}$, the greatest fixed-point of g.*
(5) *If $\boldsymbol{x}$ is such that $0 = F(\boldsymbol{x})$, then $\boldsymbol{x}_{|\mathcal{I}} \in [x^{\perp}; x^{\top}]$.*
(6) *If $x^* = x^{\perp} = x^{\top}$, then $0 = F(\boldsymbol{x})$ has a unique solution $\boldsymbol{x}$ and $\boldsymbol{x}_{|\mathcal{I}} = x^*$.*

Algorithm 1. Computation of the anti-monotonic function f underlying an anti-monotonic RN.

Require: Anti-monotonic network (V, E)
and $y \in \prod_{i \in \mathcal{I}}[0; c_i]$
1: **set** done **to** $V^- \cup \{i \in V^+ \mid \text{in}(i) = \emptyset\}$
2: **for all** $i \in$ done **do**
3:     **if** $i \in \mathcal{I}$ **then**
4:       **set** $f_i(x)$ **to** y_i
5:     **else**
6:       **set** $f_i(x)$ **to** ϕ_i
7:     **end if**
8: **end for**
9: **set** left **to** $\bigcup_{i \in \text{done}} \text{out}(i)$
10: **while** left $\neq \emptyset$ **do**
11:     **set** tmp **to** left
12:     **for all** $i \in$ tmp **do**
13:       **if** $\text{in}(i) \subseteq$ done **then**
14:         **set** $f_i(x)$ **to** $\phi_i(u^+, u^-)$, where u^+ and u^- are computed from $\{f_j(x) \mid j \in \text{in}(i)\}$
15:         **set** done **to** done $\cup \{i\}$
16:         **set** left **to** left $\cup$ (out$(i) \setminus$ done)
17:       **end if**
18:     **end for**
19: **end while**
20: **for all** $i \in \mathcal{I}$ **do**
21:     **set** $f_i(x)$ **to** $\phi_i(u^+)$, where u^+ is computed from $\{f_j(x) \mid j \in \text{in}(i)\}$
22: **end for**
23: **return** $f(x) \in \prod_{i \in V}[0; c_i]$

Proof. We start by noting that $f_{|\mathcal{I}} : \prod_{i \in \mathcal{I}}[0; c_i] \to \prod_{i \in \mathcal{I}}[0; c_i]$ is a continuous function that maps a compact set into itself (the continuity is ensured by the fact that all ϕ_i are continuous). Hence, Brouwer's fixed-point theorem ensures (1). Instead, (2) follows from Theorem 1 that ensures that f is anti-monotonic. Statement (3) and (4) are readily implied by Kleene's fixed-point theorem. To see (5) and (6), we first note that any fixed-point of $f_{|\mathcal{I}}$ is necessarily a fixed-point of g. This and Theorem 1, which establishes that the fixed-points of $f_{|\mathcal{I}}$ stand in an one-to-one correspondence with the equilibria of the dynamical system underlying the regulatory network, yield the claim.

Theorem 2 formalizes the claims made in Sect. 2 and provides an analysis framework for the equilibria of an ODE underlying an anti-monotonic RN.

We conclude the section by noting that one may be tempted to check $x^\perp = x^\top$ by solving the ODE system for different initial conditions (that respect the maximal attainable concentration vector c). However, while such an ad-hoc approach can be used to discover the presence of different equilibria, it cannot be used to prove their absence because the number of possible initial conditions is infinite.

4 Evaluation

We next discuss a model for the EGFR signaling pathway from [5], whose associated RN is given in Fig. 4a. It is also reported, with a compact set of labels, in Fig. 4b, to which we shall refer from now on.

We use a mass-action semantics that follows [6], where each vertex is associated with a variable triplet. Here we already express it in the more compact notation which exploits the preservation of mass among the three forms. Thus we use two ODE variables x_i^* and x_i, which represent the active and the passive

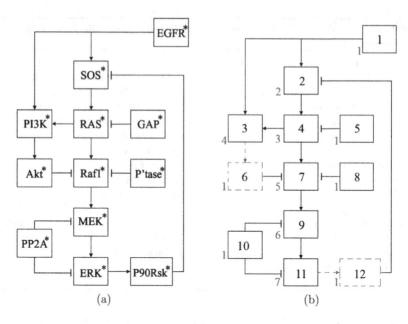

(a)   (b)

Fig. 4. Left: Regulatory network from [5] modeling an EGFR pathway. Right: The corresponding open network which arises by suppressing the activations from 3 to 6 and 11 to 12, respectively. The open network has the core observables 6 and 12. The blue numbers give the while loop iteration at whose beginning the corresponding f value becomes available for the first time. For instance, f_6 and f_{12} are initialized during the for loop in line 2 and are thus available at the beginning of the first while loop iteration. Instead, f_2 is computed during the first while loop iteration, hence it is not available at the beginning of the first while loop iteration. (Color figure online)

form of each component, respectively, and denote by c_i the total concentration for vertex i. Let us denote by $\Omega_i^+ \subseteq V$ and $\Omega_i^- \subseteq V$ the set of activators and inhibitors of vertex i, respectively. The ODEs are given by

$$\dot{x}_i^* = -\left(\sum_{I_k \in \Omega_i^-} \beta_{k,i} x_{I_k} + \sum_{A_l \in \Omega_i^+} \alpha_{k,i}(c_{A_l} - x_{A_l}) \right) x_i^*$$
$$+ \left(\sum_{A_l \in \Omega_i^+} \alpha_{k,i} x_{A_l} + \sum_{I_k \in \Omega_i^-} \beta_{k,i}(c_{I_k} - x_{I_k}) \right)(c_i - x_i^* - x_i)$$
$$\dot{x}_i = -\left(\sum_{A_l \in \Omega_i^+} \alpha_{k,i} x_{A_l} + \sum_{I_k \in \Omega_i^-} \beta_{k,i}(c_{I_k} - x_{I_k}) \right) x_i \tag{7}$$
$$+ \left(\sum_{I_k \in \Omega_i^-} \beta_{k,i} x_{I_k} + \sum_{A_l \in \Omega_i^+} \alpha_{k,i}(c_{A_l} - x_{A_l}) \right)(c_i - x_i^* - x_i)$$

where the parameters α and β are positive constants that give the strengths of inhibition and activation, respectively.

Table 1. Application of Algorithm 1 to the model of the epidermal growth factor from Fig. 4. The variable values are stated with respect to the beginning of the while loop iteration one, two, three, four and five, respectively. All function terms are meant to be evaluated with respect to x_6 and x_{12}, e.g., $f_2 \equiv f_2(x_6, x_{12})$ and $\phi_2(f_1, f_{12}) \equiv \phi_2(f_1(x_6, x_{12}), f_{12}(x_6, x_{12}))$. This is because 6 and 12 are treated as inputs during the while loop. The for loop from line 20, instead, takes the suppressed feedbacks to 6 and 12 into account and assigns $f_6 := \phi_6(f_3)$ and $f_{12} := \phi_{12}(f_{11})$

Value	Iteration 1	Iteration 2	Iteration 3	Iteration 4	Iteration 5	Iteration 6
f_1	ϕ_1	ϕ_1	ϕ_1	ϕ_1	ϕ_1	ϕ_1
f_2	—	$\phi_2(f_1, f_{12})$	$\phi_2(f_1, f_{12})$	$\phi_2(f_1, f_{12})$	$\phi_2(f_1, f_{12})$	$\phi_2(f_1, f_{12})$
f_3	—	—	—	$\phi_3(f_1, f_4)$	$\phi_3(f_1, f_4)$	$\phi_3(f_1, f_4)$
f_4	—	—	$\phi_4(f_2, f_5)$	$\phi_4(f_2, f_5)$	$\phi_4(f_2, f_5)$	$\phi_4(f_2, f_5)$
f_5	ϕ_5	ϕ_5	ϕ_5	ϕ_5	ϕ_5	ϕ_5
f_6	x_6	x_6	x_6	x_6	x_6	x_6
f_7	—	—	—	$\phi_7(f_4, f_6, f_8)$	$\phi_7(f_4, f_6, f_8)$	$\phi_7(f_4, f_6, f_8)$
f_8	ϕ_8	ϕ_8	ϕ_8	ϕ_8	ϕ_8	ϕ_8
f_9	—	—	—	—	$\phi_9(f_7, f_{10})$	$\phi_9(f_7, f_{10})$
f_{10}	ϕ_{10}	ϕ_{10}	ϕ_{10}	ϕ_{10}	ϕ_{10}	ϕ_{10}
f_{11}	—	—	—	—	—	$\phi_{11}(f_9, f_{10})$
f_{12}	x_{12}	x_{12}	x_{12}	x_{12}	x_{12}	x_{12}

The set of core observables is $\mathcal{I} = \{6, 12\}$ and conditions (i)–(iii) can be easily seen to hold true. Instead, using

$$a_i(A, I) = \sum_{A_l} \alpha_{l,i} x_{A_l} - \sum_{I_k} \beta_{k,i} x_{I_k} + \sum_{I_k} \beta_{k,i} c_{I_k}$$

$$b_i(A, I) = \sum_{I_k} \beta_{k,i} x_{I_k} - \sum_{A_l} \alpha_{l,i} x_{A_l} + \sum_{A_l} \alpha_{l,i} c_{A_l},$$

it can be shown that $\phi_i = c_i a_i^2 / (a_i^2 + a_i b_i + b_i^2)$ is the unique equilibrium point of x_i^*. Moreover, it can be proven that the roots of $(\partial_{A_l} \phi_i)(A_l)$ are either less than or equal to zero or strictly greater than c_{A_l}. By considering $A_l \mapsto (\partial_{A_l}^2 \phi_i)(A_l)$ at $A_l = 0$, we infer that ϕ_i is monotonic in A_l on $[0; c_{A_l}]$. A similar argumentation ensures that ϕ_i is anti-monotonic in every I_k on $[0; c_{I_k}]$, thus yielding the following.

Lemma 2. *The semantics (7) satisfies condition (iv).*

Proof. Additionally to the observations made above that ensure that (iv) holds true, we remark that the equilibrium of the non-observable species of vertex i (i.e., x_i) is given by $c_i b_i^2 / (a_i^2 + a_i b_i + b_i^2)$. Hence, for any fixed-point of $y \mapsto f_{|\mathcal{I}}(y)$, vector $f(y)$ and $c_i b_i^2 / (a_i^2 + a_i b_i + b_i^2)$ induce the equilibrium associated to y, see second statement of Theorem 1.

The above discussion allows us to apply Algorithm 1. The order in which it computes the entries of the vector function f are provided in Fig. 4b (light blue numbers alongside the nodes). Table 1 additionally provides the corresponding f values.

Experiments. We evaluated our approach on the MAPK and EGFR model from Figs. 1 and 4, respectively. To this end, we ran 100 experiments for each model in which the parameters were uniformly sampled from the compact interval $[1; 1000]$, covering thus a variety of possible scenarios; for both models we used the initial conditions stated in [18] and [5], respectively. In each experiment, the ODE system could be shown to enjoy a unique equilibrium because the least and the greatest fixed-point of g from Theorem 2 were identical.

5   Conclusion

In this paper we have presented an algorithm for the computation of equilibrium points in ODE systems that model regulatory networks. Our method rests on the idea of cutting feedback loops in the network, similarly to other approaches in the literature, most notably [2,13] and [25]. In contrast to [25], our approach does not require to solve systems of nonlinear equations and enjoys, under certain assumptions, polynomial time complexity. The work of Sontag et al. [2,13], instead, does not focus on the computation of equilibria. Instead, it ensures that convergence to unstable equilibria is only possible from a set of initial conditions with Lebesgue measure zero.

A possible line of future research is to combine the computational approach of the present paper with [2,13]. Specifically, the goal would be to provide sufficient conditions ensuring that the set of equilibria contains, for instance, only attractors. To this end, we plan to relate the anti-monotonic regulatory networks to the graphs that are induced by systems of differential equations in [1–3,13]. Moreover, we intend to investigate whether the requirement that each vertex is either an inhibitor or activator can be dropped by introducing artificial vertices that preserve equilibria. Finally, aiming at a more efficient computation of equilibria, we also plan to investigate if the current approach can be combined with model reduction techniques such as [7–10,12,14].

Acknowledgments. The authors thank the anonymous reviewers for helpful comments. Max Tschaikowski is supported by a Lise Meitner Fellowship funded by the Austrian Science Fund (FWF) under grant number M-2393-N32 (COCO).

References

1. Angeli, D., De Leenheer, P., Sontag, E.: Graph-theoretic characterizations of monotonicity of chemical networks in reaction coordinates. J. Math. Biol. **61**(4), 581–616 (2010)
2. Angeli, D., Ferrell, J.E., Sontag, E.D.: Detection of multistability, bifurcations, and hysteresis in a large class of biological positive-feedback systems. Proc. Natl. Acad. Sci. **101**(7), 1822–1827 (2004)

3. Angeli, D., Sontag, E.D.: Multi-stability in monotone input/output systems. Syst. Control Lett. **51**(3–4), 185–202 (2004)
4. Bianconi, F., Baldelli, E., Ludovini, V., Crinò, L., Flacco, A., Valigi, P.: Computational model of EGFR and IGF1R pathways in lung cancer: a systems biology approach for translational oncology. Biotechnol. Adv. **30**(1), 142–153 (2012)
5. Brown, K.S., et al.: The statistical mechanics of complex signaling networks: nerve growth factor signaling. Phys. Biol. **1**(3), 184 (2004)
6. Cardelli, L.: Morphisms of reaction networks that couple structure to function. BMC Syst. Biol. **8**(1), 84 (2014)
7. Cardelli, L., Tribastone, M., Tschaikowski, M., Vandin, A.: Forward and backward bisimulations for chemical reaction networks. In: CONCUR, pp. 226–239 (2015)
8. Cardelli, L., Tribastone, M., Tschaikowski, M., Vandin, A.: Comparing chemical reaction networks: a categorical and algorithmic perspective. In: LICS, pp. 485–494 (2016)
9. Cardelli, L., Tribastone, M., Tschaikowski, M., Vandin, A.: Symbolic computation of differential equivalences. In: POPL, pp. 137–150 (2016)
10. Cardelli, L., Tribastone, M., Tschaikowski, M., Vandin, A.: Maximal aggregation of polynomial dynamical systems. Proc. Natl. Acad. Sci. (PNAS) **114**(38), 10029–10034 (2017)
11. Cook, B., Fisher, J., Krepska, E., Piterman, N.: Proving stabilization of biological systems. In: Jhala, R., Schmidt, D. (eds.) VMCAI 2011. LNCS, vol. 6538, pp. 134–149. Springer, Heidelberg (2011). https://doi.org/10.1007/978-3-642-18275-4_11
12. Danos, V., Feret, J., Fontana, W., Harmer, R., Krivine, J.: Abstracting the differential semantics of rule-based models: exact and automated model reduction. In: LICS, pp. 362–381 (2010)
13. Enciso, G., Smith, H., Sontag, E.: Nonmonotone systems decomposable into monotone systems with negative feedback. J. Differ. Equ. **224**(1), 205–227 (2006)
14. Feret, J., Danos, V., Krivine, J., Harmer, R., Fontana, W.: Internal coarse-graining of molecular systems. Proc. Natl. Acad. Sci. **106**(16), 6453–6458 (2009)
15. Fussmann, G.F., Ellner, S.P., Shertzer, K.W., Hairston Jr., N.G.: Crossing the Hopf bifurcation in a live predator-prey system. Science **290**(5495), 1358–1360 (2000)
16. Gilbert, D., et al.: Computational methodologies for modelling, analysis and simulation of signalling networks. Brief. Bioinform. **7**(4), 339–353 (2006)
17. Grosu, R., et al.: From cardiac cells to genetic regulatory networks. In: Gopalakrishnan, G., Qadeer, S. (eds.) CAV 2011. LNCS, vol. 6806, pp. 396–411. Springer, Heidelberg (2011). https://doi.org/10.1007/978-3-642-22110-1_31
18. Kholodenko, B.N.: Negative feedback and ultrasensitivity can bring about oscillations in the mitogen-activated protein kinase cascades. Eur. J. Biochem. **267**(6), 1583–1588 (2000)
19. Kitano, H.: Biological robustness. Nat. Rev. Genet. **5**, 826 (2004)
20. Knoll, D., Keyes, D.: Jacobian-free Newton-Krylov methods: a survey of approaches and applications. J. Comput. Phys. **193**(2), 357–397 (2004)
21. Markevich, N.I., Hoek, J.B., Kholodenko, B.N.: Signaling switches and bistability arising from multisite phosphorylation in protein kinase cascades. J. Cell Biol. **164**(3), 353–359 (2004)
22. Orton, R.J., Adriaens, M.E., Gormand, A., Sturm, O.E., Kolch, W., Gilbert, D.R.: Computational modelling of cancerous mutations in the EGFR/ERK signalling pathway. BMC Syst. Biol. **3**(1), 100 (2009)
23. Orton, R.J., Sturm, O.E., Vyshemirsky, V., Calder, M., Gilbert, D.R., Kolch, W.: Computational modelling of the receptor-tyrosine-kinase-activated MAPK pathway. Biochem. J. **392**(2), 249–261 (2005)

24. Pappalardo, F., et al.: Computational modeling of PI3K/AKT and MAPK signaling pathways in melanoma cancer. PLoS ONE **11**(3), e0152104 (2016)
25. Radde, N.: Fixed point characterization of biological networks with complex graph topology. Bioinformatics **26**(22), 2874–2880 (2010)
26. Radde, N.: Analyzing fixed points of intracellular regulation networks with interrelated feedback topology. BMC Syst. Biol. **6**(1), 57 (2012)
27. Tyson, J.J., Novák, B.: Functional motifs in biochemical reaction networks. Annu. Rev. Phys. Chem. **61**(1), 219–240 (2010). pMID: 20055671
28. Vaudry, D., Stork, P., Lazarovici, P., Eiden, L.: Signaling pathways for PC12 cell differentiation: making the right connections. Science **296**(5573), 1648–1649 (2002)
29. Volinsky, N., Kholodenko, B.N.: Complexity of receptor tyrosine kinase signal processing. Cold Spring Harb. Perspect. Biol. **5**(8), a009043 (2013)

Rejection-Based Simulation of Stochastic Spreading Processes on Complex Networks

Gerrit Großmann$^{(\boxtimes)}$ and Verena Wolf

Saarland University, 66123 Saarbrücken, Germany
{gerrit.grossmann,verena.wolf}@uni-saarland.de
https://mosi.uni-saarland.de/

Abstract. Stochastic processes can model many emerging phenomena on networks, like the spread of computer viruses, rumors, or infectious diseases. Understanding the dynamics of such stochastic spreading processes is therefore of fundamental interest. In this work we consider the wide-spread compartment model where each node is in one of several states (or compartments). Nodes change their state randomly after an exponentially distributed waiting time and according to a given set of rules. For networks of realistic size, even the generation of only a single stochastic trajectory of a spreading process is computationally very expensive.

Here, we propose a novel simulation approach, which combines the advantages of event-based simulation and rejection sampling. Our method outperforms state-of-the-art methods in terms of absolute runtime and scales significantly better while being statistically equivalent.

Keywords: Spreading process · SIR · Epidemic modeling · Monte-Carlo simulation · Gillespie Algorithm

1 Introduction

Computational modeling of spreading phenomena is an active research field within network science with many applications ranging from disease prevention to social network analysis [1–6]. The most widely used approach is a continuous-time model where each node of a given graph occupies one of several states (e.g. *infected* and *susceptible*) at each point in time. A set of rules determines the probabilities and random times at which nodes change their state depending on the node's direct neighborhood (as determined by the graph). The application of a rule is always stochastic and the waiting time before a rule "fires" (i.e. is applied) is governed by an exponential distribution.

The underlying stochastic dynamics are given by a continuous-time Markov chain (CTMC) [6–9]. Each possible assignment from nodes to local node states constitutes an individual state of the CTMC (here referred to as *CTMC state* or *network state* to avoid confusion with the local state of a single node). Hence, the

© Springer Nature Switzerland AG 2019
M. Češka and N. Paoletti (Eds.): HSB 2019, LNBI 11705, pp. 63–79, 2019.
https://doi.org/10.1007/978-3-030-28042-0_5

corresponding CTMC state space grows exponentially in the number of nodes, which renders its numerical solution infeasible.

As a consequence, mean-field-type approximations and sampling approaches have emerged as the cornerstones for their analysis. Mean-field equations originate from statistical physics and provide typically a reasonably good approximation of the underlying dynamics [10–14]. Generally speaking, they propose a set of ordinary differential equations that model the average behavior of each component (e.g., for each node, or for all nodes of a certain degree). However, mean-field approaches only give information about the *average behavior* of the system, for example, about the expected number of *infected* nodes for each degree. Naturally, this restricts the scope of their application. In particular, they are not suited to answer specific questions about the system.

For example, one might be interested in finding the specific source of an epidemic [15,16] or wants to know where an intervention (e.g. by vaccination) is most successful [17–20].

Consequently, stochastic simulations remain an essential tool in the computational analysis of complex networks dynamics. Different simulation approaches for complex networks have been suggested, which can all be seen as adaptations of the *Gillespie algorithm* (GA) [6]. Recently, a more efficient extension of the GA has been proposed, called *Optimized GA* (OGA) [21]. A rejection step is used to reduce the number of network updates.

Here, we propose an event-driven simulation method which also utilizes rejection sampling. Our method is based on an event queue which stores infection and curing events. Unlike traditional methods, we ensure that it is not necessary to iterate over the entire neighborhood of a node after it has changed its state. Therefore, we allow the creation of events which are inconsistent with the current CMTC state. These might lead to rejections when they reach the beginning of the queue. We introduce our method for the well-known SIS (Susceptible-Infected-Susceptible) model and show that it can easily be generalized for other epidemic-type processes. Code will be made available.[1]

We formalize the semantics of spreading processes in Sect. 2 and explain how the CTMC is constructed. Previous simulation approaches, such as GA and OGA, are presented in Sect. 3. In Sect. 4 we present our rejection sampling algorithm and discuss to which extend our method is generalizable to different network models and spreading models. We demonstrate the effectiveness of our approach on three different case studies in Sect. 5.

2   Stochastic Spreading Processes

Let $\mathcal{G} = (\mathcal{N}, \mathcal{E})$ be a an undirected, unweighted, finite graph without self-loops. We assume the edges are tuples of nodes and that $(n_1, n_2) \in \mathcal{E}$ always implies $(n_2, n_1) \in \mathcal{E}$. At each time point $t \in \mathbb{R}_{\geq 0}$ each node occupies one out of m (local) states (also called labels or compartments), denoted by $\mathcal{S} = \{s_1, s_2, \ldots, s_m\}$.

---

[1] github.com/gerritgr/Rejection-Based-Epidemic-Simulation.

Consequently, the (global) network state is fully specified by a labeling $L : \mathcal{N} \rightarrow \mathcal{S}$. We use $\mathcal{L} = \{L \mid L : \mathcal{N} \rightarrow \mathcal{S}\}$ to denote all possible network states. As each of the $|\mathcal{N}|$ nodes occupies one of m states, we know that $|\mathcal{L}| = m^{|\mathcal{N}|}$. Nodes change their state by the application of a stochastic rule. A node's state and its neighborhood determine which rules are applicable to a node and the probability density of the random delay until a rule fires. If several rules can fire, the one with the shortest delay is executed.

We allow two types of rules: *node-based (independent, spontaneous)* rules and *edge-based (contact, spreading)* rules. The application of a node-based rule $A \xrightarrow{\mu} B$ results in a transition of a node from state $A \in \mathcal{S}$ to state $B \in \mathcal{S}$ ($A \neq B$) with rate $\mu \in \mathbb{R}_{>0}$. That is, the waiting time until the rule fires is governed by the exponential distribution with rate μ. An edge-based rule has the form $A + C \xrightarrow{\lambda} B + C$, where $A, B, C \in \mathcal{S}, A \neq B, \lambda \in \mathbb{R}_{>0}$. Its application changes an edge (more precisely, the state of an edge's node). It can be applied to each edge $(n, n') \in \mathcal{E}$ where $L(n) = A, L(n') = B$. Again, the node in state A changes after a delay that is exponentially distributed with rate λ. Note that, if a node in state A has more than one direct B-neighbor, it is "attacked" independently by each neighbor. Due to the properties of the exponential distribution, the rate at which a node changes its state according to a certain contact rule is proportional to the number of neighbors which induce the change.

SIS Model. In the sequel, we use the well-known Susceptible-Infected-Susceptible (SIS) model as a running example. Consider $\mathcal{S} = \{I, S\}$ and the rules:

$$S + I \xrightarrow{\lambda} I + I \qquad I \xrightarrow{\mu} S.$$

In the SIS model, infected nodes propagate their infection to neighboring susceptible nodes using an edge-based rule. Thus, only susceptible nodes with at least one infected neighbor can become infected. Infection of a node occurs at a rate that increases proportionally with the number of infected neighbors. Infected nodes can, independently from their neighborhood, recover (i.e. become susceptible again) using a node-based rule.

3 Previous Approaches

In this section, we shortly revise techniques that have been previously suggested for the simulation of SIS-type processes. For a more comprehensive description, we refer the reader to [6,21].

3.1 Standard Gillespie Algorithm

The *Standard Gillespie Algorithm* (here, simply referred to as *GA*) is also known as Gillespie's direct method and a popular method for the simulation of coupled chemical reactions. Its adaptation to complex networks uses as key data structures two lists which are constantly updated: a list of all infected nodes (denoted by $\mathcal{L}_I$) and a list of all S–I edges (denoted by $\mathcal{L}_{S-I}$).

In each simulation step, we first draw an exponentially distributed delay for the time until the next rule fires. That is, instead of sampling a waiting time for each rule and each position where the rule can be applied, we directly sample the time until the network state changes. For this, we compute an aggregated rate $c = \mu|\mathcal{L}_I| + \lambda|\mathcal{L}_{S-I}|$. Then we randomly decide if an infection or a curing event is happening. The probability of the latter is proportional to its rate, i.e. $\frac{1}{c}\mu|\mathcal{L}_I|$, and thus, the probability of an infection is $\frac{1}{c}\lambda|\mathcal{L}_{S-I}|$. After that we pick an infected node (in case of a curing) or an S–I edge (in case of an infection) uniformly at random. We update the two lists accordingly. The expensive part in each step is keeping $\mathcal{L}_{S-I}$ updated. For this, we iterate over the whole neighborhood of the node and for each susceptible neighbor we remove (after a curing) or add (after an infection) the corresponding edge to the list. Thus, we need one add/remove operation on the list for each susceptible neighbor.

Note that there are different possibilities to sample the node that will become infected next. Instead of keeping an updated list of all S–I edges one can also use a list of all susceptible nodes. In that case, we cannot sample uniformly but decide for the infection of a susceptible node with a probability proportional to its number of infected neighbors.

Likewise, we can randomly pick the starting point of the next infection by only considering $\mathcal{L}_I$. To generate an infection event, we first sample an infected node from this list and then we (uniformly) sample a susceptible neighbor, which becomes infected. Since infected nodes with many susceptible neighbors have a higher probability of being the starting point of an infection (i.e., they have more S–I edges associated with them), we sample from $\mathcal{L}_I$ such that the probability of picking an infected node is proportional to its number of susceptible neighbors.

All three approaches are statistically equivalent but the last one motivates the *Optimized Gillespie Algorithm* (OGA) [21].

3.2 Optimized Gillespie Algorithm

As discussed earlier, sampling from $\mathcal{L}_I$ is expensive because Updating this information for all elements of $\mathcal{L}_I$ is costly because after each event, the number of susceptible neighbors may change for many nodes.

In [21] Cota and Ferreira suggest to sample nodes from $\mathcal{L}_I$ with a probability that is proportional to the degree k of a node, which is an upper bound for the maximal possible number of susceptible neighbors. Then they uniformly choose a neighbor of that node and update the global clock. If this neighbor is already infected they reject the infection event, which yields a rejection probability of $\frac{k-k_S}{k}$ if k_S is the number of susceptible neighbors. Note that the rejection probability exactly corrects for the over-approximation of using k instead of k_S. This is illustrated in Fig. 1.

Compared to the GA, updating the list of infected nodes becomes cheaper, because only the node which actually changes its state is added to (or removed

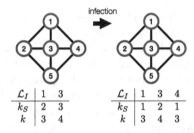

Fig. 1. Example of an infection event. We sample from $\mathcal{L}_I$ proportional to k_S. Alternatively, we can weight according to the number of neighbors k which is constant and over-approximates k_S. To correct for the over-approximation we reject a sample with probability $\frac{k-k_s}{k}$.

from) $\mathcal{L}_I$. The sampling probabilities of the neighbors remain the same because their degree remains the same. On the other hand, sampling of a node is more expensive compared to the GA where we sample edges uniformly.

Naturally, the speedup in each step comes at the costs (of a potentially enormous amount) of rejection events. Even a single infected node with many infected but few susceptible neighbors will continuously lead to rejected events. This is especially problematic in cases with many infected nodes and no or very few susceptible neighbors which therefore make rejections many orders of magnitude more likely than actual events. Therefore, in [21] the authors propose the algorithm for simulations close to the epidemic threshold, where the number of infected nodes is typically very small.

Note that, to sample a node, Cota and Ferreira also propose rejection sampling based on the maximal degree. However, St-Onge et al. point out that in the case of heterogeneous networks a binary tree can be used to speed up this step significantly. Specifically, this allows them to derive an upper-bound for the rejecting probability [22]. This, however, does not overcome the fundamental limitation of the OGA approach regarding models with a large fraction of infected nodes. That is, where infected nodes are mostly surrounded by infected nodes causing most infections to be rejected.

3.3 Event-Based Simulation

In the event-driven approach, the primary data structure is an event queue, in which events are sorted and executed according to the time points at which they will occur. This eliminates the costly process of randomly selecting a node for each step (popping the first element from the queue has constant time complexity). Events are either curing of a specific node or infection via a specific edge. Moreover, it is easy to adapt the event-driven approach to rules with non-Markovian waiting times or to a network where each node has individual recovery and infection rates [6]. Event-based simulation of an SIS process is done as follows: For the initialization, we draw for each node an exponentially distributed time until recovery with rate μ and add the respective curing event to the queue. Likewise, for each susceptible node with at least one infected neighbor we draw an exponentially distributed time until infection with rate $\lambda \times$ "Number of infected neighbors". We add the resulting events to the queue.

During the simulation, we always take the earliest event from the queue, change the network accordingly and update the global clock. If the current event is the infection of a node, the infection rates of its susceptible neighbors increase. Thus, it is necessary to iterate over all neighbors of the corresponding node, draw renewed waiting times for their infection events, and update the event queue accordingly. Although efficient strategies have been suggested [6], these queue updates are rather costly.

Since each step requires an iteration over all neighbors of the node under consideration, the worst-case runtime depends on the maximal degree of the network. Moreover, for each neighbor, it might be necessary to reorder the event queue. The time complexity of reordering the queue depends (typically logarithmically) on the number of elements in the queue and adds significant additional costs to each step. Note that trajectories generated using the event-driven approach are statistically equivalent to those generated with the GA because all delays are exponentially distributed and thus have the memoryless property. A variant of this algorithm can also be found in [23].

4   Our Method

In this section, we propose a method for the simulation of SIS-type processes. The key idea is to combine an event-driven approach with rejection sampling while keeping the number of rejections to a minimum. We will generalize the algorithm for different epidemic processes as well as for weighted and temporal networks. First, we introduce the main data structures:

Event Queue. It stores all future infection and curing events generated so far. Each event is associated with a time point and with the node(s) affected by the event. Curing events contain a reference to the recovering node and infection events to a pair of connected nodes, an infected (source) node and a susceptible (target) node.

Graph. In this graph structure, each node is associated with its list of neighbors, its current state, a degree, and, if infected, a prospective recovery time.

We also keep track of the time in a global clock. We assume that an initial network, a time horizon (or another stopping criterion), and the rate parameters (μ, λ) are given as input. In Algorithm 1–4 we provide pseudocode for the detailed steps of the method.

Initialization. Initially, we iterate over the network and sample a recovery time (exponentially distributed with rate μ) for each infected node (cf. Line 2, Algorithm 1). We push the recovery event to the queue and annotate each infected node with its recovery time (cf. Line 5, Algorithm 2). Next, we iterate over the network a second time and generate an infection event for each infected node (cf. Line 5, Algorithm 1). The procedure for the generation of infection events is explained later.

In Algorithm 1 we need two iterations because the recovery time of each infected node has to be available for the infection events. These events identify the earliest infection attempt of each node.

Iteration. The main procedure of the simulation is illustrated in Algorithm 4. We schedule events until the global clock reaches the specified time horizon (cf. Line 9). In each step, we take the earliest event from the queue (Line 7) and set the global clock to the event time (Line 8). Then we "apply" the event (Line 11–20).

In case of a recovery event, we simply change the state of the corresponding node from I to S and are done (Line 12). Note that we always generate (exactly) one recovery event for each infected node, thus, each recovery event is always consistent with the current network state. Note that the queue always contains exactly one recovery event for each infected node.

If the event is an infection event, we apply the event if possible (Line 14–18) and reject it otherwise (Line 19–20). We update the global clock either way. Each infection event is associated with a source node and a target node (i.e., the node under attack). The infection event is applicable if the current state of the target node is S (which might not be the case anymore) and the current state of the source node is I (which will always be the case). After a successful infection event, we generate a new recovery event for the target node (Line 16) and two novel infection events, one for the source node (Line 17) and one for the target node which is now also infected (Line 18). If the infection attempt was rejected, we only generate a novel infection event for the source node (Line 20). Thus, we always have exactly one infection event in the queue for each infected node.

Generating Infection Events. The generation of infection events and the distinction between unsuccessful and potentially successful infection attempts is an essential part of the algorithm.

In Algorithm 3, for each infected node we only generate the earliest infection attempt and add it to the queue. Therefore, we first sample the exponentially distributed waiting time with rate $k\lambda$, where k is the degree of the node, and compute the time point of infection (Line 5). If the time point of the infection attempt is after its recovery event, we stop and no infection event is added to the queue (Lines 6–7). Note that in the graph structure, each node is annotated with its recovery time (node.recovery_time) to have it immediately available.

Next, we uniformly select a random neighbor which will be attacked (Line 8). If the neighbor is currently susceptible, we add the event to the event queue and the current iteration step ends (Lines 9–12).

If the neighbor is currently infected we check the recovery time of the neighbor (Line 9). If the infection attempt happens before the recovery time point, we already know that the infection attempt will be unsuccessful (already infected

---

Algorithm 1 Graph Initialization

1: **procedure** INITGRAPH(G, μ, λ, Q)
2:     **for each** node in G **do**
3:         **if** node.state = I **then**
4:             GENERATERECOVERYEVENT(node, μ, 0, Q)
5:     **for each** node in G **do**   ▷ recovery times are available now
6:         **if** node.state = I **then**
7:             GENERATEINFECTIONEVENT(node, λ, 0, Q)

Algorithm 2 Generation of a Recovery Event

1: **procedure** GENERATERECOVERYEVENT(node, μ, t_{global}, Q)
2:     $t_{\text{event}} = t_{\text{global}} + \text{draw\_exp}(\mu)$
3:     e = Event(src_node = node, t=t_{event}, type=recovery)
4:     node.recovery_time = t_{event}
5:     Q.push(e)

Algorithm 3 Generation of an Infection Event

1: **procedure** GENERATEINFECTIONEVENT(node, λ, t_{global}, Q)
2:     $t_{\text{event}} = t_{\text{global}}$
3:     rate = λ*node.degree
4:     **while** true **do**
5:         t_{event} += draw_exp(rate)
6:         **if** node.recovery_time < t_{event} **then**   ▷ no event is generated
7:             break
8:         attacked_node = draw_uniform(node.neighbor_list)
9:         **if** attacked_node.state = S
            or attacked_node.recovery_time < t_{event} **then**   ▷ check for early reject
10:             e = Event(src_node=node, target=attacked_node,
                    time=t_{event}, type=infection)
11:             Q.push(e)   ▷ was successful
12:             break

Algorithm 4 SIS Simulation

    Input: Graph (G) with initial states, time horizon (h), recovery rate (μ), infection rate (λ)
    Output: Graph at time h ▷ or any other measure of interest
1: Q = EMPTYQUEUE()   ▷ sorted w.r.t. time
2: INITGRAPH(G, μ, λ, Q)
3: $t_{\text{global}} = 0$
4: **while** true **do**
5:     **if** Q.is_empty() **then**
6:         break
7:     e = Q.pop()
8:     t_{global} = e.time
9:     **if** $t_{\text{global}} > h$ **then**
10:         break
11:     **if** e.type = recovery **then**
12:         G[e.src_node].state = S
13:     **else**
14:         **if** G[e.target_node].state = S **then**
15:             G[e.target_node].state = I
16:             GENERATERECOVERYEVENT(e.target_node, μ, t_{global}, Q)
17:             GENERATEINFECTIONEVENT(e.src_node, λ, t_{global}, Q)
18:             GENERATEINFECTIONEVENT(e.target_node, λ, t_{global}, Q)
19:         **else**   ▷ late reject
20:             GENERATEINFECTIONEVENT(e.src_node, λ, t_{global}, Q)

Fig. 2. Pseudocode for our event-based rejection sampling method.

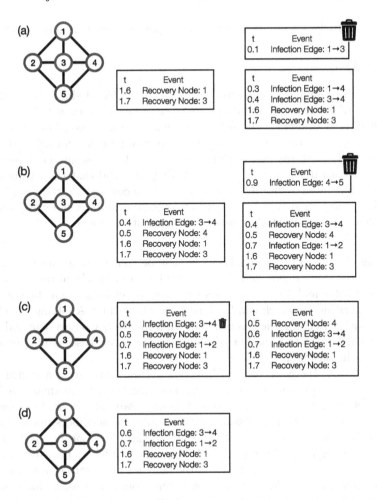

Fig. 3. First four steps of the our method for a toy example (I: red, S: blue): (a) Initialization, generate the recovery events (left queue), and infection event for each infected node (right queue). The first infection attempt from node 1 is an early reject. (b) The infection from 1 to 4 was successful, we generate a recovery event for 4 and two new infection events for 1 and 4. The infection event of node 4 is directly rejected because it happens after its recovery. (c) (Late) Reject of the infection attempt from 3 to 4 as 4 is already infected. A new infection event starting from 3 is inserted into the queue. (d) Node 4 recovers, the remaining queue is shown. (Color figure online)

nodes cannot become infected). Thus, we perform an *early reject* (Lines 10–12 are not executed). That is, instead of pushing the surely unsuccessful infection event to the queue, we directly generate another infection attempt, i.e. we re-enter the while-loop in Lines 4–12. We repeat the above procedure until the recovery time of the current node is reached or the infection can be added to the queue (i.e. no early rejection is happening) (Fig. 2).

Figure 3 provides a minimal example of a potential execution of our method.

4.1 Analysis

Our approach combines the advantages of an event-based simulation with the advantages of rejection sampling. In contrast to the *Optimized Gillespie Algorithm*, finding the node for the next event can be done in constant time. More importantly, the number of rejection events is dramatically minimized because the queue only contains events that are realistically possible. Therefore, it is crucial that each node "knows" its own curing time and that the curing events are always generated before the infection events. In contrast to traditional event-based simulation, we do not have to iterate over all neighbors of a newly infected node followed by a potentially costly reordering of the queue.

Runtime. For the runtime analysis, we assume that a binary heap is used to implement the event queue and that the graph structure is implemented using a hashmap. Each simulation step starts by taking an element from the queue (cf. Line 7, Algorithm 4), which can be done in constant time. Applying the change of state to a particular node has constant time complexity on average and linear time complexity (in the number of nodes) in the worst case as it is based on lookups in the hashmap.

Now consider the generation of infection events. Generating a waiting time (Line 3, Algorithm 3) can be done in constant time because we know the degree (and therefore the rate) of each node. Likewise, sampling a random neighbor (Line 8) is constant in time (assuming the number of neighbors fits in an integer). Checking for an *early reject* (Line 9) can also be done in constant time because each neighbor is sampled with the same (uniform) probability and is annotated with its recovery time. Even though each early rejection can be computed in constant time, the number of early rejections can of course increase with the mean (and maximal) degree of the network. Inserting the newly generated infection event(s) to the event queue (Line 11) has a worst-case time complexity of $\mathcal{O}(\log n)$, where n is the number of elements in the heap. In our case, n is bounded by twice the number of infected nodes. However, we can expect constant insertion costs on average [24, 25].

Correctness. Here, we argue that our method generates correct sample trajectories of the underlying Markov model. To see this, we assume some hypothetical changes to our method that do not change the sampled trajectories but makes it easier to reason about the correctness. First, assume that we abandon *early rejects* and insert all events in the event queue regardless of their possibility of success. Second, assume that we change the generation of infection events such that we do not only generate the earliest attempt but all infection attempts until recovery of the node. Note that we do not do this in practice, as this would lead to more rejections (less early rejections).

Similar to [21], we find that our algorithm is equivalent to the direct event-based implementation of the following spreading process:

$$ I \xrightarrow{\mu} S \qquad S + I \xrightarrow{\lambda} I + I \qquad I + I \xrightarrow{\lambda} I + I . $$

In [21], $I + I \xrightarrow{\lambda} I + I$ is called a shadow process, because the application of this rule does not change the network state. Hence, rejections of infections in the SIS model can be interpreted as applications of the shadow process. Note that the rate at which this rule is applied to the network is the rate of the rejection events. Hence, the rate at which an infected node attacks its neighbors (no matter whether in state I or S) is exactly λk, where k is the degree of the node. Our method includes the shadow process into our simulation in the following way: For each S–I edge and I–I edge, an infection event is generated with rate λ and inserted into the queue. The decision if this event will be a real or a "shadow infection" is postponed until the event is actually applied. This is possible because both rules have the same rate, in particular, the joint rate at which an infected k-degree node attacks its neighbors will always be $k\lambda$.

4.2 Generalizations

So far we have only considered SIS processes on static and unweighted networks. This section shorty discusses how to generalize our simulation method to SIS-type processes on temporal and weighted networks.

General Epidemic Models. A key ingredient to our algorithm is the early rejection of infection events. This is possible because we can compute a node's curing time already when the node gets infected. In particular, we exploit that there is only one way to leave state I, that is, by the application of a node-based rule. This gives us a guarantee about the remaining time in state I. Other epidemic models have a similar structure. For instance, consider the Susceptible-Infected-Recovered (SIR) model, where infected nodes first become recovered (immune), before entering state I again:

$$ S + I \xrightarrow{\lambda} I + I \qquad I \xrightarrow{\mu_1} R \qquad R \xrightarrow{\mu_2} S . $$

We also consider the competing pathogens model [26], where two infectious diseases, denoted by I and J, compete over the susceptible nodes:

$$ S + I \xrightarrow{\lambda_1} I + I \qquad S + J \xrightarrow{\lambda_2} J + J \qquad I \xrightarrow{\mu_1} S \qquad J \xrightarrow{\mu_2} S . $$

In both cases, we can exploit that certain states (I, J, R) can only be left under node-based rules and thus their residence time is independent of their neighborhood. This makes it simple to annotate each node in any of these states with their exact residence time and perform early rejections accordingly.

If we do not have these guarantees, early rejection cannot be applied. For instance in the (fictional) system:

$$S + I \xrightarrow{\lambda_1} I + I \qquad\qquad I + I \xrightarrow{\lambda_2} I + S .$$

It is likely that our method will still perform better than the traditional event-based approach, however, the number of rejection events might significantly decrease its performance.

Weighted Networks. In weighted networks, each edge $e \in \mathcal{E}$ is associated with a positive real-valued weight $w(e) \in \mathbb{R}_{>0}$. Each edge-based rule of the form

$$A + C \xrightarrow{\lambda} B + C$$

fires on this particular edge with rate $w(e) \cdot \lambda$. Applying our method to weighted networks is simple: Let n be a node. During the generation of infection events, instead of sampling the waiting time with rate λk, we now use $\lambda \sum_{n' \in N(n)} w(n, n')$ as the rate, where $N(n)$ is the set of neighbors of n. Moreover, instead of choosing a neighbor that will be attacked with uniform probability, we choose them with a probability proportionally to their edge weight. This can be done by rejection sampling or in $\mathcal{O}(\log(k))$ time complexity, where k is the degree of n.

Temporal Networks. Temporal (time-varying, adaptive, dynamic) networks are an intriguing generalization of static networks which generally complicates the analysis of their spreading behavior [27–30]. Generalizing the Gillespie algorithm for Markovian epidemic-type processes is far from trivial [27].

In order to keep our model as general as possible, we assume here that an external process governs the temporal changes in the network. This process runs simultaneously to our simulation and might or might not depend on the current network state. It changes the current graph by adding or removing edges, one edge at a time. For instance, after processing one event, the external process could add or remove an arbitrary number of edges at specific time points until the time of the next event is reached. It is simple to integrate this into our simulation.

Given that the external process removes an edge, we can simply update the neighbor list and the degrees in our graph. For each infection event that reaches the top of the queue, we first check if the corresponding edge is still present. If not, we reject the event. This is possible because removing events only decreases infection rates which we can correct by using rejections. When an edge is added to the graph and at least one corresponding node is infected, the infection rate increases. Thus, it is not sufficient to only update the graph, we also generate an infection event which accounts for the new edge. In order to minimize the number of generated events, we change the algorithm such that each infected node is annotated with the time point of its subsequent infection attempt. Consider now an infected node. When it obtains a new edge, we generate an exponentially distributed waiting time with rate λ modeling the infection attempt through this specific link. We only generate a new event if this time point lies before the time point of the subsequent infection attempt of the node. In that case, we also remove the old event associated with this node from the queue.

Since most changes in the graph do not require changes to the event queue (and those that do only cause two operations at maximum), we expect our method to handle temporal networks with a reasonably high number of graph updates very efficiently. In the case that an extremely large number of edges in the graph change at once, we can always decide to iterate over the whole network and newly initialize the event queue.

5 Case Studies

We demonstrate the effectiveness our approach on three classical epidemic-type processes. We compare the performance of our method with the *Standard Gillespie Algorithm* (GA) and the *Optimized Gillespie Algorithm* (OGA) for different network sizes. We use synthetically generated networks following the configuration model [31] with a truncated power-law degree distribution, that is $P(k) \propto k^{-\gamma}$ for $3 \le k \le 1000$. We compare the performance on degree distributions with $\gamma \in \{2, 3\}$. This yields a mean degree around 30 ($\gamma = 2$) and 10 ($\gamma = 3$). We use models from the literature but adapt rate parameters freely to generate interesting dynamics. Nevertheless, we find that our observations generalize to a wide range of parameters that yield networks with realistic degree distributions and spreading dynamics.

We also report how the number of nodes in a network is related to the CPU time of a single step. This is more informative than using the total runtime of a simulation because the number of steps obviously increases with the number of nodes when the time horizon is fixed. The CPU time per step is defined as the total runtime of the simulation divided by the number of steps, only counting the steps that actually change the network state (i.e., excluding rejections). We do not count rejection events, because that would give an unfair advantage to the rejection based approach. The evaluation was performed on a 2017 MacBook Pro with a 3.1 GHz Intel Core i5 CPU and 16 GB of RAM.

Note that an implementation of the OGA was only available for the SIS model and the comparison is therefore not available for other models. Due to the high number of rejection steps in all models, we expect a similar difference in the performance between our approach and the OGA also for other models.

5.1 SIS Model

For the SIS model we used rate parameters of $(\mu, \lambda) = (1.0, 0.6)$ and an initial distribution of 95% susceptible nodes and 5% infected nodes. CPU times are reported in Fig. 4a, where "reject" refers to our rejection-based algorithm (as described in Sect. 4). For a sample trajectory, we plot the fraction of nodes in

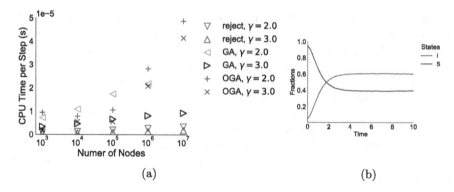

Fig. 4. SIS model (a): Average CPU time for a single step (i.e., change of network state) for different networks. The GA method run out of memory for $\gamma = 2.0$, $|N| = 10^7$. (b): Sample dynamics for a network with $\gamma = 3.0$ and 10^5 nodes.

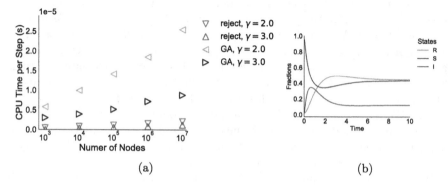

Fig. 5. SIR model (a): Average CPU time for a single step (i.e., change of network state) for different networks. (b): Sample dynamics for a network with $\gamma = 2.0$ and 10^5 nodes.

each state w.r.t. time (Fig. 4b). To have a comparison with OGA we used the official Fortran-implementation in [21] and estimated the average CPU time per step based on the absolute runtime. Note that the comparison is not perfectly fair due to implementation differences and additional input/output of the OGA code. It is not surprising that the OGA performs comparably bad, as the method is suited for simulations close to the epidemic threshold. Moreover, our maximal degree is very large, which negatively affects the performance of the OGA.

We also conducted experiments on models closer to the epidemic threshold (i.e., where the number of infection events is very small, e.g. $\lambda = 0.1$) and with smaller maximal degree (e.g. $k_{max} = 100$). The relative speed-up to the GA increased slightly compared to the results in Fig. 4a. The performance of the OGA improved significantly compared our method leading to a similar performance as our method (results not shown).

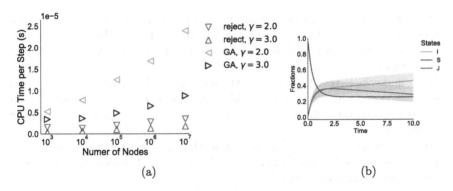

Fig. 6. Competing pathogens model (a): Average CPU time for a single step (i.e., change of network state) for different networks. (b): Mean fractions and standard deviations of a network with $\gamma = 2.0$ and 10^4 nodes.

5.2 SIR Model

Next, we considered the SIR model, which has more complex dynamics. We used rate parameters of $(\mu_1, \mu_2, \lambda) = (1.1, 0.3, 0.6)$ and an initial distribution of 96% susceptible nodes and 2% infected and recovered nodes, respectively. Similar as above, CPU times and example dynamics are reported in Fig. 5. We see that runtime behavior is almost the same as in the SIS model.

5.3 Competing Pathogens Model

Finally, we considered the Competing Pathogens model. We used rate parameters of $(\lambda_1, \lambda_2, \mu_1, \mu_2) = (0.6, 0.63, 0.6, 0.7)$ and an initial distribution of 96% susceptible nodes and 2% infected nodes for both pathogens (denoted by I, J), respectively. CPU times and network dynamics are reported in Fig. 6. The model is interesting because we see that in the beginning J dominates I due to its higher infection rate. However, nodes infected with pathogen J recover faster than those infected with I. This gives the I pathogen the advantage that infected nodes have more time to attack their neighbors. In the limit, I takes over and J dies out. For this model stochastic noise has a significant influence on the macroscopic dynamics. Therefore, we also reported the standard deviation of the fractions (cf. Fig. 6). Note that the fraction of susceptible nodes is almost deterministic. Performance-wise our rejection method performs slightly worse than in the previous models (w.r.t. the baseline). We believe that this is due to the even larger number of infection events and rejections.

6 Conclusions

In this paper, we presented a novel rejection algorithm for the simulation of epidemic-type processes. We combined the advantages of rejection sampling and event-driven simulation. In particular, we exploited that nodes can only leave certain states using node-based rules, which made it possible to precompute their residence times, which then again allowed us to perform early rejection of certain events.

Our numerical results show that our method outperforms previous approaches especially for networks which are not close the epidemic threshold. In particular, the speed-up increases as the maximal degree of the network increases.

As future work, we plan to extend the method to compartment models with arbitrary rules, including an automated decision for which states early rejections can be computed and are useful.

Acknowledgments. This research has been partially funded by the German Research Council (DFG) as part of the Collaborative Research Center "Methods and Tools for Understanding and Controlling Privacy". We thank Michael Backenköhler for his comments on the manuscript.

References

1. Barabási, A.-L.: Network Science. Cambridge University Press, Cambridge (2016)
2. Barrat, A., Barthelemy, M., Vespignani, A.: Dynamical Processes on Complex Networks. Cambridge University Press, Cambridge (2008)
3. Porter, M., Gleeson, J.: Dynamical Systems on Networks: A Tutorial, vol. 4. Springer, Heidelberg (2016). https://doi.org/10.1007/978-3-319-26641-1
4. Goutsias, J., Jenkinson, G.: Markovian dynamics on complex reaction networks. Phys. Rep. **529**(2), 199–264 (2013)
5. Pastor-Satorras, R., Castellano, C., Van Mieghem, P., Vespignani, A.: Epidemic processes in complex networks. Rev. Mod. Phys. **87**(3), 925 (2015)
6. Kiss, I.Z., Miller, J.C., Simon, P.L.: Mathematics of Epidemics on Networks: From Exact to Approximate Models. IAM, vol. 46. Springer, Cham (2017). https://doi.org/10.1007/978-3-319-50806-1
7. Simon, P.L., Taylor, M., Kiss, I.Z.: Exact epidemic models on graphs using graph-automorphism driven lumping. J. Math. Biol. **62**(4), 479–508 (2011)
8. Van Mieghem, P., Omic, J., Kooij, R.: Virus spread in networks. IEEE/ACM Trans. Netw. (TON) **17**(1), 1–14 (2009)
9. Sahneh, F.D., Scoglio, C., Van Mieghem, P.: Generalized epidemic mean-field model for spreading processes over multilayer complex networks. IEEE/ACM Trans. Netw. (TON) **21**(5), 1609–1620 (2013)
10. Gleeson, J.P.: High-accuracy approximation of binary-state dynamics on networks. Phys. Rev. Lett. **107**(6), 068701 (2011)
11. Gleeson, J.P., Melnik, S., Ward, J.A., Porter, M.A., Mucha, P.J.: Accuracy of mean-field theory for dynamics on real-world networks. Phys. Rev. E **85**(2), 026106 (2012)
12. Gleeson, J.P.: Binary-state dynamics on complex networks: pair approximation and beyond. Phys. Rev. X **3**(2), 021004 (2013)
13. Devriendt, K., Van Mieghem, P.: Unified mean-field framework for susceptible-infected-susceptible epidemics on networks, based on graph partitioning and the isoperimetric inequality. Phys. Rev. E **96**(5), 052314 (2017)

14. Bortolussi, L., Hillston, J., Latella, D., Massink, M.: Continuous approximation of collective system behaviour: a tutorial. Perform. Eval. **70**(5), 317–349 (2013)
15. Prakash, B.A., Vreeken, J., Faloutsos, C.: Spotting culprits in epidemics: how many and which ones? In: 2012 IEEE 12th International Conference on Data Mining (ICDM), pp. 11–20. IEEE (2012)
16. Farajtabar, M., Gomez-Rodriguez, M., Du, N., Zamani, M., Zha, H., Song, L.: Back to the past: source identification in diffusion networks from partially observed cascades. In: Artificial Intelligence and Statistics (2015)
17. Schneider, C.M., Mihaljev, T., Havlin, S., Herrmann, H.J.: Suppressing epidemics with a limited amount of immunization units. Phys. Rev. E **84**(6), 061911 (2011)
18. Cohen, R., Havlin, S., Ben-Avraham, D.: Efficient immunization strategies for computer networks and populations. Phys. Rev. Lett. **91**(24), 247901 (2003)
19. Buono, C., Braunstein, L.A.: Immunization strategy for epidemic spreading on multilayer networks. EPL (Europhys. Lett.) **109**(2), 26001 (2015)
20. Wu, Q., Fu, X., Jin, Z., Small, M.: Influence of dynamic immunization on epidemic spreading in networks. Phys. A **419**, 566–574 (2015)
21. Cota, W., Ferreira, S.C.: Optimized Gillespie algorithms for the simulation of Markovian epidemic processes on large and heterogeneous networks. Comput. Phys. Commun. **219**, 303–312 (2017)
22. St-Onge, G., Young, J.-G., Hébert-Dufresne, L., Dubé, L.J.: Efficient sampling of spreading processes on complex networks using a composition and rejection algorithm. arXiv preprint arXiv:1808.05859 (2018)
23. Sahneh, F.D., Vajdi, A., Shakeri, H., Fan, F., Scoglio, C.: GEMFsim: a stochastic simulator for the generalized epidemic modeling framework. J. Comput. Sci. **22**, 36–44 (2017)
24. Hayward, R., McDiarmid, C.: Average case analysis of heap building by repeated insertion. J. Algorithms **12**(1), 126–153 (1991)
25. Porter, T., Simon, I.: Random insertion into a priority queue structure. IEEE Trans. Softw. Eng. **SE–1**(3), 292–298 (1975)
26. Masuda, N., Konno, N.: Multi-state epidemic processes on complex networks. J. Theor. Biol. **243**(1), 64–75 (2006)
27. Vestergaard, C.L., Génois, M.: Temporal gillespie algorithm: fast simulation of contagion processes on time-varying networks. PLoS Comput. Biol. **11**(10), e1004579 (2015)
28. Masuda, N., Holme, P.: Temporal Network Epidemiology. Springer, Heidelberg (2017). https://doi.org/10.1007/978-981-10-5287-3
29. Holme, P., Saramäki, J.: Temporal networks. Phys. Rep. **519**(3), 97–125 (2012)
30. Holme, P.: Modern temporal network theory: a colloquium. Eur. Phys. J. B **88**(9), 234 (2015)
31. Fosdick, B.K., Larremore, D.B., Nishimura, J., Ugander, J.: Configuring random graph models with fixed degree sequences. SIAM Rev. **60**(2), 315–355 (2018)

Controlling Noisy Expression Through Auto Regulation of Burst Frequency and Protein Stability

Pavol Bokes[1(✉)] and Abhyudai Singh[2]

[1] Department of Applied Mathematics and Statistics, Comenius University,
84248 Bratislava, Slovakia
`pavol.bokes@fmph.uniba.sk`
[2] Department of Electrical and Computer Engineering, University of Delaware,
Newark, DE 19716, USA
`absingh@udel.edu`

Abstract. Protein levels can be controlled by regulating protein synthesis or half life. The aim of this paper is to investigate how introducing feedback in burst frequency or protein decay rate affects the stochastic distribution of protein level. Using a tractable hybrid mathematical framework, we show that the two feedback pathways lead to the same mean and noise predictions in the small-noise regime. Away from the small-noise regime, feedback in decay rate outperforms feedback in burst frequency in terms of noise control. The difference is particularly conspicuous in the strong-feedback regime. We also formulate a fine-grained discrete model which reduces to the hybrid model in the large system-size limit. We show how to approximate the discrete protein copy-number distribution and its Fano factor using hybrid theory. We also demonstrate that the hybrid model reduces to an ordinary differential equation in the limit of small noise. Our study thus contains a comparative evaluation of feedback in burst frequency and decay rate, and provides additional results on model reduction and approximation.

1 Introduction

Synthesis of protein molecules in bursts of multiple copies has been identified as a major factor in gene expression noise [13]. The number of bursts per protein lifespan determines the abundance of a bursty protein [10]. This ratio can be controlled by the numerator, the burst frequency, or the denominator, the protein decay rate. Feedback in burst frequency has been widely documented [2], and examples of feedback in decay rate are available too [16,27]. Linear noise approximation based analysis suggest that the two feedback pathways are equivalent in terms of controlling gene-expression noise [26].

In this paper we compare the two feedback pathways using a hybrid model for bursty gene expression with negative feedback in burst frequency or decay rate. Hybrid models mix continuous deterministic with discrete stochastic dynamics [11,12,19,23]. The chosen modelling framework is hybrid in that it combines

© Springer Nature Switzerland AG 2019
M. Češka and N. Paoletti (Eds.): HSB 2019, LNBI 11705, pp. 80–97, 2019.
https://doi.org/10.1007/978-3-030-28042-0_6

stochastic dynamics of bursty production occurring at discrete time-points with deterministic dynamics of protein decay [4].

Intuition suggests that, by repressing production, negative feedback lowers the protein mean, and, by improving regression to the mean, it also lowers the protein noise [14]. Counter-intuitively, multiple studies report that the response of noise to increasing feedback strength is U-shaped [20, 28]. The eventual increase in the noise can be attributed e.g. to low copy number effects [25], loss of time averaging [18], or the failure to control large bursts [5]. In this paper we examine how the choice of feedback pathway (burst frequency or decay rate) affects the shape of the noise response to strengthening feedback.

The outline of the paper is as follows. Section 2 introduces the chosen hybrid modelling framework on a protein which is expressed constitutively without feedback. Section 3 extends the hybrid model by negative autoregulation, and Sect. 4 derives the steady-state distribution for the extended model. Section 5 defines a specific noise metric that is used here to evaluate feedback performance. Sections 6 and 7 elaborate on feedback in burst size and decay rate, respectively, the two feedback pathways whose performance we are set to compare. Section 8 introduces a full discrete model for bursty protein expression. Section 9 contains the bulk of the results of this paper that are based on the theoretical backbone of the previous sections. The results compare the performance of the two feedback types, and draw connections between the full discrete, the hybrid and the deterministic modelling formalisms. Section 10 concludes the paper with a short summary.

2   Constitutive Model

By the constitutive model we understand a hybrid stochastic bursting gene-expression model without a feedback mechanism. The protein level dynamics is given by the balance of deterministic protein decay and stochastic protein synthesis in bursts. Between bursts, the protein concentration satisfies a linear ordinary differential equation $dx/dt = -\gamma x$, where γ is the decay rate constant, implying that the temporal profile of protein concentration is piecewise exponential (see Fig. 1, left). Bursts occur randomly in time with frequency α per unit time. It follows that the waiting time from one burst until the next one is drawn from the exponential distribution with mean waiting time $1/\alpha$. The size of a burst is also random and is drawn from the exponential distribution with mean burst size β [10].

The probability balance equation (i.e. the master equation) for the hybrid process as described above takes the form of a partial integro-differential equation [24]

$$\frac{\partial p}{\partial t} + \frac{\partial J}{\partial x} = 0, \quad J = -\gamma x p(x, t) + \alpha \int_0^x \exp\left(-\frac{x-y}{\beta}\right) p(y, t) dy. \quad (1)$$

The solution $p(x, t)$ gives the probability density function of protein concentration x at time t. The partial integro-differential equation (1) represents a hybrid analogue of the chemical master equation used in discrete systems [7].

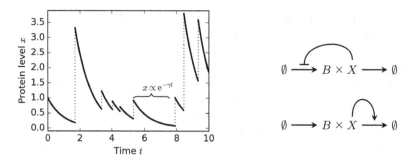

Fig. 1. *Left:* Sample protein path corresponding to the constitutive model (1). The concentration of protein decays with rate constant γ between bursts (solid lines). Bursts occur randomly in time with burst frequency α and lead to positive discontinuous jumps of mean size β in protein level (dotted vertical lines). In this example, $\alpha = \beta = \gamma = 1$. *Right:* The two feedback types considered in this paper are feedback in burst frequency and feedback in decay rate. Protein X is produced in bursts of size B and degrades one molecule at a time. The protein controls its level either by reducing the frequency of burst occurrence or by enhancing its own decay.

The first equation in (1) states the principle of probability conservation in differential form. It says that the probability changes in time due to differentials in the probability flux J. The probability flux is described in the second equation of (1). The flux consists of a negative local flux and a positive non-local flux. In general, a flux is local if it depends on the value of the solution at the point x of flux evaluation, whereas a non-local flux depends on the values of the solution away from the evaluation point; the sign of a flux corresponds to the direction of probability mass transfer. In our particular model (1), the negative local flux represents the downward transfer of probability mass due to deterministic decay of protein, and the positive non-local flux represents the upward transfer of probability mass due to bursts of protein production. Note in particular that the integral kernel in the nonlocal flux expresses the probability that a burst occurs which takes the protein concentration from a value y below x into any value above x.

Previous studies established that the gamma distribution [15]

$$p(x) = \frac{1}{\Gamma(a)\beta^a}x^{a-1}e^{-\frac{x}{\beta}} \tag{2}$$

is a steady-state solution to the master Eq. (1). The parameter a in (2) is defined by

$$a = \frac{\alpha}{\gamma}, \tag{3}$$

and gives the average number of protein bursts per protein lifetime. The mean and variance of (2) are

$$\langle x \rangle = a\beta, \quad \mathrm{Var}(x) = a\beta^2. \tag{4}$$

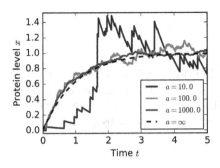
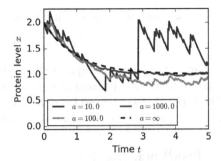

Fig. 2. As the average number a of bursts per protein lifetime increases, the trajectories of the stochastic bursting model (6) become well approximated by the solution (9) to the ordinary differential equation model (8). Without loss of generality, the burst size is scaled to $1/a$ in order that the steady-state protein mean is equal to 1 regardless of the choice of a.

It follows immediately from (4) that the squared coefficient of variation, defined as the ratio of the variance to the square of mean, is equal to a^{-1}. Therefore, if a large number of bursts occur on average per protein lifetime, the noise in protein concentration is low.

It is convenient to measure the protein concentration in units of its mean and time in units of the protein lifetime. This is achieved via nondimensionalisation

$$x = a\beta\tilde{x}, \quad t = \frac{\tilde{t}}{\gamma}, \quad p(x,t) = \tilde{p}(\tilde{x},\tilde{t}), \tag{5}$$

where $\tilde{x}$ and $\tilde{t}$ represent the dimensionless concentration and time variables. We insert (5) into (1) and, for simplicity, drop the tildes in the dimensionless variables symbols, obtaining

$$\frac{\partial p}{\partial t} + \frac{\partial J}{\partial x} = 0, \quad J = -xp(x,t) + a \int_0^x \exp(-a(x-y))p(y,t)dy. \tag{6}$$

Comparing the dimensional problem (1) to the dimensionless problem (6), we see that the latter can be formally obtained from the former by setting $\alpha = a$, $\beta = 1/a$, $\gamma = 1$. However, the assignment $\beta = 1/a$ should not be interpreted as implying that burst sizes are physically small if the burst frequency is large. Rather, it means that burst sizes are small in comparison to the steady-state mean.

In the regime $a \gg 1$ of very frequent (and very short) bursts, the probability flux J in (6) can be approximated using the Laplace method [17] by

$$J \sim (1-x)p(x,t) \quad \text{for } a \gg 1. \tag{7}$$

The reduced flux (7) corresponds to deterministic dynamics governed by the ordinary differential equation

$$\dot{x} = 1 - x, \tag{8}$$

whose solutions have the explicit form of

$$x(t) = 1 + (x(0) - 1)e^{-t}. \tag{9}$$

Figure 2 compares sample paths of the hybrid model (6) to the deterministic solution (9) for two different initial conditions, $x(0) = 0$ (Fig. 2, left) and $x(0) = 2$ (Fig. 2, right). As expected from the use of the Laplace method, stochastic sample paths are close to the deterministic solution for large burst frequencies a.

3   Feedback Model

Here we extend the hybrid stochastic model (6) with feedback in burst frequency and decay rate (Fig. 1, right). In the feedback model, the probability of a burst to occur in a time interval of length dt is equal to $ah(x)dt + o(dt)$, where x gives the current protein concentration and $h(x)$ is a response function as specified below. Bursts sizes are exponentially distributed with mean size $1/a$ like in the (dimensionless) constitutive model. Between bursts, the protein concentration satisfies $\dot{x} = -g(x)x$, in which $g(x)$ is another response function. We assume that the response functions satisfy

$$g(0) = 1, \quad g'(x) \geq 0, \quad h(0) = 1, \quad h'(x) \leq 0. \tag{10}$$

If there is a shortage of protein (x is close to zero), bursts occur with frequency a and decay with rate constant 1 as in the constitutive model. However, as the protein concentration increases, bursts become less frequent and/or the propensity of protein molecules for decay increases.

The master equation for the feedback model reads

$$\frac{\partial p}{\partial t} + \frac{\partial J}{\partial x} = 0, \quad J = -g(x)xp(x,t) + a \int_0^x \exp(-a(x-y))h(y)p(y,t)dy. \tag{11}$$

Applying the Laplace method on the non-local flux yields

$$J \sim (h(x) - g(x)x)p(x,t) \quad \text{for } a \gg 1, \tag{12}$$

which corresponds to the ordinary differential equation

$$\dot{x} = h(x) - g(x)x. \tag{13}$$

Under assumptions (10), Eq. (13) has a single globally stable steady state which is smaller than the steady state 1 of the constitutive deterministic model (8).

4   Steady State Distribution and Moments

At steady state, the probability flux J in the master Eq. (11) vanishes, leading to the Volterra integral equation

$$g(x)xp(x) = a \int_0^x \exp(-a(x-y))h(y)p(y)dy \tag{14}$$

for the stationary protein distribution $p(x)$. In order to solve (14) in the unknown $p(x)$, we multiply both sides by e^{ax},

$$e^{ax}g(x)xp(x) = a \int_0^x e^{ay}h(y)p(y)\mathrm{d}y, \tag{15}$$

and differentiate with respect to x to obtain a linear first-order ordinary differential equation

$$\frac{\mathrm{d}}{\mathrm{d}x}\left(e^{ax}g(x)xp(x)\right) = ae^{ax}h(x)p(x) = \frac{ah(x)}{xg(x)} \times e^{ax}g(x)xp(x). \tag{16}$$

Solving (16) in $e^{ax}g(x)xp(x)$ implies that up to a normalisation constant we have

$$p(x) = \frac{e^{a\Phi(x)}}{xg(x)}, \tag{17}$$

in which the potential $\Phi(x)$ is defined through the indefinite integral

$$\Phi(x) = \int \frac{h(x)}{xg(x)}\mathrm{d}x - x. \tag{18}$$

The n-th moment of the steady-state protein distribution is given by

$$\langle x^n \rangle = \frac{M_n}{M_0}, \quad \text{where } M_n = \int_0^\infty x^n p(x)\mathrm{d}x. \tag{19}$$

In general, the moments can be evaluated by numerical integration of (19). In special parametric regimes, asymptotic approximations to the integrals (19) can be developed (Appendices A and B). In the next section, we use the moments (19) to define a specific characteristic of protein noise.

5 Relative Noise

In this section we provide a definition of relative noise in protein concentration. The purpose of this quantity is to compare the steady-state variance in a feedback model to the steady-state variance in a referential constitutive model. The latter is chosen so as to have the same steady-state mean as the feedback model. By doing such a comparison, we compensate for the increase in noise in the feedback model that results from the decrease of the time-averaged number of bursts per protein lifetime. What remains is the change in noise that results from improved mean reversion in a feedback model. Indeed, we shall see that the relative noise is always less than 1 in our examples of negative autoregulatory pathways. For this section only, we refer to the concentration of a self-regulating protein as x_{reg} and to the concentration of the referential constitutive protein as x_{const}.

In the absence of regulation, the normalised burst frequency is equal to a and the burst size is equal to $1/a$. These values lead to the mean value of 1.

In order to satisfy the constraint $\langle x_{\text{const}} \rangle = \langle x_{\text{reg}} \rangle$, we decrease the normalised burst frequency in the constitutive model to $a\langle x_{\text{reg}} \rangle$ whilst keeping the burst size equal to $1/a$. Since the protein variance is equal to the product of burst frequency and the square of burst size, cf. (4), we find that

$$\text{Var}(x_{\text{const}}) = \frac{\langle x_{\text{reg}} \rangle}{a}. \tag{20}$$

The relative noise compares the variances in the regulated model and the referential constitutive model,

$$\eta^2 = \frac{\text{Var}(x_{\text{reg}})}{\text{Var}(x_{\text{const}})} = a\frac{\text{Var}(x_{\text{reg}})}{\langle x_{\text{reg}} \rangle} = a\left(\frac{M_2}{M_1} - \frac{M_1}{M_0}\right). \tag{21}$$

The definition (21) of the relative noise superficially resembles the Fano factor [6]. However, the two should not be confused. The value one of Fano factor means Poissonian noise. On the other hand, $\eta^2 = 1$ means that the regulated protein has the same variance as the referential unregulated protein. Nevertheless, that can still correspond to a very large Fano factor: how large the actual Fano factor is depends on how many molecule copies are encompassed in an average burst. In Sect. 8, we consider a discrete modelling approach and systematically establish the relationship between the Fano factor of a full discrete model and the relative noise of the hybrid model.

6 Feedback in Burst Frequency

Sections 3–4 provided general results for feedback in burst frequency and decay rate acting concurrently. Here we provide additional details for the situation if feedback is in burst frequency only. We thereby focus on a specific type of response function, the decreasing Hill function. This leads to choices

$$h(x) = \frac{1}{1 + (x/K)^H}, \quad g(x) = 1 \tag{22}$$

in the general model (11). The parameter K gives the critical concentration of protein that is required to halve the burst frequency. The parameter H is the cooperativity coefficient. Large values of H imply that the burst frequency decreases rapidly from its maximal value to zero as the protein concentration exceeds the critical threshold K. The critical threshold K is a reciprocal measure of feedback strength: small values of K mean that low amounts of protein suffice to turn off the production. For this reason we refer from now on to the reciprocal K^{-1} of the critical threshold as feedback strength. It is easy to verify that the choices in (22) satisfy the assumptions (10) imposed on the feedback model.

With choices (22), the limiting ordinary differential equation (13) takes the form of

$$\dot{x} = \frac{1}{1 + (x/K)^H} - x. \tag{23}$$

Solutions to (23) converge to a unique, globally stable, steady state, which satisfies the fixed-point equation

$$\frac{1}{1 + (x_0/K)^H} = x_0. \tag{24}$$

Elementary analysis shows that x_0 is an increasing function of K, i.e. that the deterministic steady state x_0 decreases with increasing feedback strength K^{-1}.

With choices (22), the potential (18) is an elementary function

$$\Phi(x) = \int \frac{dx}{x(1 + (x/K)^H)} - x = \ln x - \frac{\ln\left(1 + (x/K)^H\right)}{H} - x. \tag{25}$$

Inserting (25) into (17) we find an explicit formula

$$p(x) = e^{-ax} x^{a-1} \left(1 + (x/K)^H\right)^{-\frac{a}{H}} \tag{26}$$

for the steady-state protein pdf which holds up to a normalisation constant. The asymptotic behaviour of the mean $\langle x \rangle$ (19) and the relative noise η^2 (21) for the protein pdf (26) in the small-noise regime ($a \gg 1$) and the strong feedback regime ($K \ll 1$) is provided in Appendices A and B.

7   Feedback in Decay Rate

Here we explore in detail the situation if feedback is in decay rate only. Specifically, we use the choices

$$h(x) = 1, \quad g(x) = 1 + (x/K)^H. \tag{27}$$

The polynomial response function $g(x)$ consist of a basal term 1 and a monomial term which is proportional to x^H. Biologically, this means that in addition to spontaneous decay, there is an additional decay pathway, which is cooperatively activated by the protein itself. The critical concentration K gives the amount of protein that is necessary to double the rate of decay per protein molecule. Small values of K mean that few proteins suffice to turn on the decay, suggesting that the reciprocal K^{-1} can again be used as a measure of feedback strength.

With choices (27), the limiting ordinary differential equation (13) reads

$$\dot{x} = 1 - (1 + (x/K)^H)x. \tag{28}$$

Equation (28) describes a different time-dependent dynamics than the limiting Eq. (23) for feedback in burst frequency. Nevertheless, solutions to (28) converge to the same steady-solution x_0 satisfying (24). Furthermore, the probability-distribution potential (18) for the choices (27) is the same as (25) obtained for feedback in burst frequency.

The steady-state protein pdf (17) simplifies to

$$p(x) = e^{-ax} x^{a-1} \left(1 + (x/K)^H\right)^{-\frac{a}{H}-1}. \tag{29}$$

We note that the pdf (29) differs from (26) only in the exponent of the third factor. The asymptotic behaviour of the mean $\langle x \rangle$ (19) and the relative noise η^2 (21) for the protein pdf (29) in the small-noise regime ($a \gg 1$) and the strong feedback regime ($K \ll 1$) is provided in Appendices A and B.

8   Discrete Approach

The full discrete model for feedback in burst frequency and decay rate is based on chemical reactions

$$\emptyset \xrightarrow{ah\left(\frac{P}{\Omega}\right)} B \times P, \quad P \xrightarrow{g\left(\frac{P}{\Omega}\right)} \emptyset. \tag{30}$$

The first reaction in (30) is the production of protein P in bursts of size B. The second reaction in (30) is the degradation of protein P. The response functions depend on the ratio P/Ω of the protein copy number to a system-size parameter Ω. Large values of Ω mean that feedback is sensitive to large changes in protein molecules. As with the hybrid model, we treat the discrete model (30) separately for the choices (22) (feedback in burst frequency) and the choices (27) (feedback in decay rate).

The burst size B is assumed to be drawn from the geometric distribution [22] with mean Ω/a

$$\text{Prob}[B = n] = \frac{a}{a + \Omega} \left(\frac{\Omega}{a + \Omega} \right)^n, \quad n = 0, 1, 2 \ldots \tag{31}$$

Due to previously developed theoretical arguments [3,9,21], the protein concentration $x = P/\Omega$ approximately satisfies in the large system-size limit $\Omega \gg 1$ the hybrid bursting model (11).

The Fano factor, which is defined as the variance to mean ratio, is a widely used measure of variability in discrete probability distributions and discrete stochastic models for gene expression [6]. The protein Fano factor satisfies

$$F = \frac{\text{Var}(P)}{\langle P \rangle} = \Omega \frac{\text{Var}(x)}{\langle x \rangle} \sim \frac{\eta^2 \Omega}{a}, \quad \text{for } \Omega \gg 1, \tag{32}$$

where η^2 is the relative noise of the hybrid model as defined by (21). Hence, for large system sizes, the Fano factor is proportional to the mean burst size Ω/a, with the relative noise of the hybrid model giving the factor of proportionality.

9   Results

This paper explores a bursting model for stochastic gene expression with negative feedback. The model is hybrid in the sense that it combines a deterministic decay of protein with stochastic protein production in bursts. Two separate versions of the model are considered, depending on whether the feedback is in burst frequency or decay rate. The model (either version) is characterised by three parameters: the maximal burst frequency a; the critical concentration K; and the cooperativity coefficient H.

The maximal burst frequency a gives the average number of bursts per protein lifetime at full gene activation. Without loss of generality, the mean burst size is scaled to $1/a$. This choice of scaling implies that the mean protein level, which is equal to the product of the burst frequency and the burst size, is bounded by one. In the regime $a \gg 1$ of frequent bursts, the trajectories of the stochastic model fluctuate near the deterministic solution (Fig. 2). The regime $a \gg 1$ is therefore referred to as the small-noise regime.

The parameters K and H determine the character of the feedback response. The critical concentration K gives the amount of protein that is required to halve the frequency of bursts (in case of feedback in burst frequency) or double the propensity for decay (in case of feedback in decay rate). The reciprocal K^{-1} is used as a measure of feedback strength. The regime $K \ll 1$ (i.e. $K^{-1} \gg 1$) is referred to as the strong-feedback regime of the model. The cooperativity coefficient H determines how steeply the response changes as the protein concentration passes through the critical threshold K.

Figure 3 shows the steady-state values of protein mean and relative noise as functions of feedback strength. The relative noise is defined in Eq. (21) as the ratio of the variance of the protein with feedback to the variance of a constitutively expressed protein with the same mean. Several values of the maximal burst frequency a are selected, including the limit value of $a \to \infty$, the results for which are derived in Appendix A using a small-noise approximation. The cooperativity coefficient is set to $H = 4$.

The small-noise approximation leads to the same mean and noise values for both feedback types. In particular, the small-noise approximation suggests that, regardless of the feedback type, a maximal $(H+1)$-fold reduction of relative noise can be achieved in the limit $K \to 0$ of strong feedback. However, the assumption of high burst frequency, on which the use of small-noise approximation is based, eventually breaks down as feedback strengthens. Indeed, finite values of the maximal burst frequency a paint a radically different picture from that obtained by the small-noise approximation. In case of feedback in burst frequency, the relative noise starts increasing for large feedback strengths, eventually returning to the value of one. Contrastingly, in case of feedback in decay rate, the relative noise decreases down to zero. Thus, despite the small-noise prediction that the two feedbacks are indistinguishable in terms of controlling protein mean and noise, we see that at high feedback strengths, feedback in decay rate can be much more effective than feedback in burst frequency.

The stark differences between the small-noise prediction and the exact results motivate us to develop in Appendix B an alternative asymptotic approximation in the regime $K \ll 1$ of strong feedback. For feedback in decay rate the asymptotics (B6) reveal that the relative noise is: of the order of K if $H > 2$; of the (asymptotically larger) order of K^{H-1} if $1 < H < 2$; or converges to the constant $1 - H$ if $0 < H < 1$. For feedback in burst frequency the strong-feedback asymp-

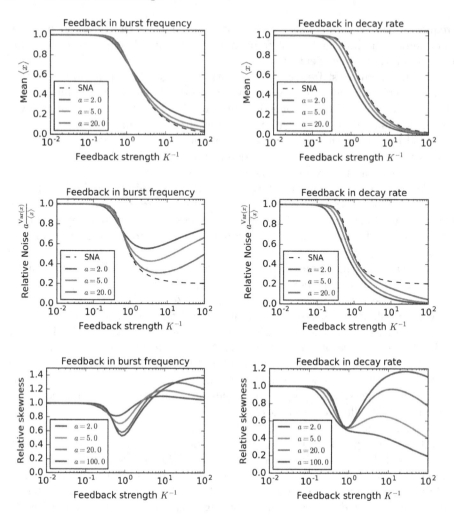

Fig. 3. The mean, the relative noise and skewness of protein concentration subject to feedback in burst frequency or decay rate. The exact values (coloured lines) are based on numerical integration of (19), in which the probability density function $p(x)$ is given by (26) (feedback in burst frequency) or (29) (feedback in decay rate). The small-noise approximation (SNA) of the protein mean is the fixed-point solution x_0 to Equation (24); the SNA of the relative noise is given by (A4). The feedback cooperativity coefficient is fixed to $H = 4$ throughout. (Color figure online)

totics [8] confirm the numerical observation that the relative noise eventually returns to the value of one as K tends to zero.

The protein skewness is quantified by the third standardised moment of its steady-state distribution. In the nethermost panels of Fig. 3 we report the response of a relative protein skewness to increasing feedback strength. By the relative skewness we understand the ratio of the skewness of the auto-regulated

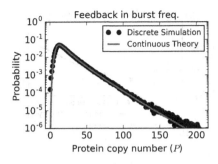
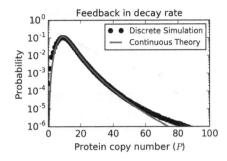

Fig. 4. Steady-state protein copy number distribution by discrete simulations and hybrid (continuous) theory. The theoretical distribution is given by $\frac{1}{M_0 \Omega} p\left(\frac{P}{\Omega}\right)$, where p is the continuous probability density function (26) (feedback in burst frequency) or (29) (feedback in decay rate) of the hybrid model and M_0 is the zero-th moment (19) (the reciprocal of the normalisation constant). Discrete simulation results are based on 10^6 Gillespie iterations of the discrete model (30), in which the response functions $h(x)$ and $g(x)$ are given by (22) (feedback in burst frequency) or (27) (feedback in decay rate). The model parameters are: burst frequency $a = 5$; cooperativity coefficient $H = 4$; critical concentration $K = 0.1$; system size $\Omega = 100$.

protein to that of a referential constitutive protein with the same mean. The protein skewness responds to increasing feedback strength in a complicated manner featuring first a trough and then a peak. Feedback in burst frequency is more conducive to skewness than feedback in decay rate.

The hybrid framework is cross-validated by a fine-grained discrete stochastic bursting model (30) with feedback in burst frequency or decay rate. In the discrete model, burst sizes are geometrically distributed; decay is stochastic and leads to the removal of one molecule at a time. In addition to the three parameters of the hybrid model, the discrete model features an additional system-size parameter Ω, which is equal to the copy number P of protein that are encompassed in a unit of protein concentration x. Provided that Ω is large, discrete protein distributions obtained by stochastic simulation of the discrete model (30) are well approximated by the explicit continuous protein distributions (26) or (29) obtained using hybrid theory (Fig. 4).

The variability of a discrete distribution is conveniently quantified using the Fano factor (the variance to mean ratio). Figure 5 shows the Fano factor for the steady-state protein copy number obtained by stochastic simulation of the full discrete model (30) and the hybrid theory approximation (32). For large values of Ω the two agree well. For small values of Ω, single-molecule effects become important in discrete simulations; by neglecting them, the hybrid theory tends to underestimate the Fano factor.

The hybrid approximation (32) implies that the protein Fano factor is proportional to the relative noise of the hybrid model. The factor of proportionality is the mean copy number Ω/a of protein molecules produced per burst. The hybrid model can thus be consistent with a range of different Fano factors depending on

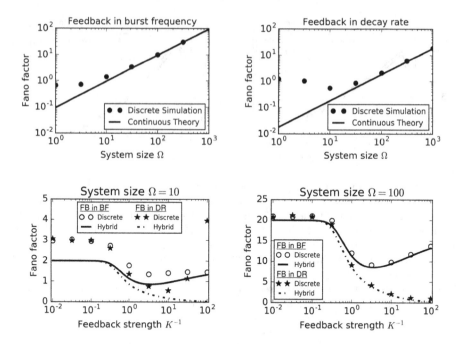

Fig. 5. The Fano factor of steady-state protein copy number as function of system size Ω (upper panels) and feedback strength K^{-1} (lower panels). The discrete results are obtained by simulation of system (30). The hybrid (continuous) are based on (32), (21), and (19). The model parameters are set to $a = 5$, $H = 4$, $K = 0.1$ (upper panels) $\Omega = 10$ or $\Omega = 100$ (lower panels). The number of Gillespie iterations is set to $10^4 \times \lfloor \Omega \rfloor$.

the chosen value of the system size. Provided that the value of the system size is fixed to a sufficiently large value, the response of the Fano factor to increasing feedback strength coincides with that of the relative noise (Fig. 5, lower panels).

10 Summary

We evaluated the noise suppression capabilities of feedback in burst frequency and feedback in decay rate using a hybrid model for bursty protein dynamics. Using a relative noise measure, we systematically related the noise levels of a regulated protein to those of a constitutive protein expressed at the same mean value. It was found that introducing feedback of either kind brings about a decrease in the relative noise. Nevertheless, feedback in decay rate was shown to perform better in suppressing noise, in particular under high-noise and/or strong-feedback conditions.

We identified the relationships between the hybrid model and other modelling frameworks, in particular a deterministic one, based on an ordinary differential equation, and a discrete stochastic framework. The deterministic model is recovered from the hybrid model in the limit of very frequent bursts. The discrete stochastic model reduces to the hybrid model in the limit of large system sizes. Discrete protein distributions estimated by a kinetic Monte Carlo method were found to be in agreement with the continuous distributions provided explicitly by the hybrid framework. The relative noise metric from the hybrid framework was shown to determine the leading order behaviour of the protein Fano factor in the large system size regime.

Overall, our results illustrate the tractability and usefulness of hybrid frameworks in studying non-linear fluctuations in stochastic gene expression.

Acknowledgement. PB is supported by the Slovak Research and Development Agency under the contract No. APVV-14-0378 and by the VEGA grant 1/0347/18. AS is supported by the National Science Foundation grant ECCS-1711548.

Appendix A: Small-Noise Asymptotics

The potential (25), which applies for both regulation types (22) and (27), is a concave function with a single maximum situated at $x = x_0$, where x_0 satisfies the fixed-point equation (24).

For $a \gg 1$, the most important part of the pdf lies around the maximum $x = x_0$ of the potential. Therefore we use the parabolic approximation in (17) to obtain

$$p(x) = \frac{Ce^{a\Phi(x)}}{xg(x)} \approx \frac{Ce^{a\left(\Phi(x_0) + \frac{\Phi''(x_0)}{2}(x-x_0)^2\right)}}{x_0 g(x_0)} = C'e^{\frac{a\Phi''(x_0)}{2}(x-x_0)^2}, \qquad (A1)$$

where $C' = Ce^{a\Phi(x_0)}/x_0 g(x_0)$ is a constant. The parabolic approximation (A1) implies that at steady state the protein concentration is normally distributed with statistics

$$\langle x \rangle \sim x_0, \quad \mathrm{Var}(x) \sim -\frac{1}{a\Phi''(x_0)}. \qquad (A2)$$

Note that the variance in (A2) is in fact positive since the second derivative of the potential $\Phi(x)$ at the point $x = x_0$ of its maximum is negative.

We see that the (leading-order) approximations (A2) to the protein statistics in the small-noise regime depend on their shared potential (25) but not on the fine differences between the pdfs (26) and (29). Therefore we arrive at a first important conclusion of the present work: feedback in burst frequency and feedback in decay rate are equivalent in terms of control of both mean and noise in the small-noise regime.

Evaluating the second derivative of the potential yields

$$\Phi''(x_0) = \frac{d}{dx} \frac{1}{x(1 + (x/K)^H)}\bigg|_{x=x_0} = -\frac{1 + (H+1)(x/K)^H}{x^2(1 + (x/K)^H)^2}\bigg|_{x=x_0}$$

$$= -\left(1 + (H+1)\left(\frac{1}{x_0} - 1\right)\right) = -\frac{H(1 - x_0) + 1}{x_0}. \qquad (A3)$$

in which we used the fixed-point Eq. (24) several times. For the relative noise (21) we find

$$\eta^2 = a\frac{\mathrm{Var}(x)}{\langle x \rangle} \sim -\frac{1}{x_0 \Phi''(x_0)} = \frac{1}{H(1-x_0)+1}. \tag{A4}$$

The asymptotic approximation of the relative noise on left-hand side of (A4), which we denote by η_{SNA}^2, is a decreasing function of K which satisfies

$$\eta_{\mathrm{SNA}}^2 \sim \frac{1}{H+1} \quad \text{for } K \ll 1. \tag{A5}$$

The small-noise asymptotics thus predict that a maximal $(H+1)$-fold reduction in relative noise can be achieved in the strong-feedback regime using either feedback type.

Appendix B: Strong-Feedback Asymptotics

In this Section we develop the relative noise asymptotics for the strong feedback regime $K \ll 1$. We separately treat feedback in decay rate and refer to literature [8] for treatment of feedback in burst frequency.

B.1 Feedback in Decay Rate

The purpose of this section is to provide asymptotic approximations as $K \ll 1$ to the integral

$$M_n = \int_0^\infty e^{-ax} x^{a+n-1} \left(1 + (x/K)^H\right)^{-\frac{a}{H}-1} \mathrm{d}x, \tag{B1}$$

giving the n-th moment of the protein pdf (29). In particular, M_0^{-1} gives the normalisation constant C in the protein pdf.

If $K \ll 1$ and $x = O(1)$, then $x/K \gg 1$ so that

$$\left(1 + (x/K)^H\right)^{-\frac{a}{H}-1} \sim (x/K)^{-a-H} \quad \text{for } K \ll 1. \tag{B2}$$

Inserting (B2) into (B1) we find

$$M_n \sim K^{a+H} \int_0^\infty e^{-ax} x^{n-H-1} \mathrm{d}x = K^{a+H} a^{H-n} \Gamma(n-H), \tag{B3}$$

which converges for $n > H$. For $n < H$, we need to use a different method of approximating the integral (B1).

Substituting $(x/K)^H = z$ in the integral (B1) yields

$$M_n = \frac{K^{a+n}}{H} \int_0^\infty e^{-aKz^{\frac{1}{H}}} z^{\frac{a+n}{H}-1}(1+z)^{-\frac{a}{H}-1}\mathrm{d}z. \tag{B4}$$

Neglecting the $O(K)$ term in the exponential in (B4) yields

$$
\begin{aligned}
M_n &\sim \frac{K^{a+n}}{H} \int_0^\infty z^{\frac{a+n}{H}-1}(1+z)^{-\frac{a}{H}-1} dz \\
&= \frac{K^{a+n}}{H} B\left(\frac{a+n}{H}, 1 - \frac{n}{H}\right), \quad \text{for } K \ll 1,
\end{aligned}
\tag{B5}
$$

where $B(\mu, \nu)$ is the beta function [1]. The right-hand side in (B5) converges for $n < H$, which complements the condition for validity of the previous approximation (B5). The nongeneric case $n = H$ can be treated by method of splitting the integration range [17].

Using the asymptotic approximations (B3) and (B5) in the formula $\eta^2 = a(M_2/M_1 - M_1/M_0)$ for the relative noise, we obtain asymptotic approximations

$$
\eta^2 \sim \begin{cases}
aK \left(\dfrac{B\left(\frac{a+2}{H}, 1 - \frac{2}{H}\right)}{B\left(\frac{a+1}{H}, 1 - \frac{1}{H}\right)} - \dfrac{B\left(\frac{a+1}{H}, 1 - \frac{1}{H}\right)}{B\left(\frac{a}{H}, 1\right)} \right) & \text{if } H > 2, \\[4mm]
\dfrac{a^{H-1} H K^{H-1} \Gamma(2 - H)}{B\left(\frac{a+1}{H}, 1 - \frac{1}{H}\right)} & \text{if } 1 < H < 2, \\[4mm]
1 - H & \text{if } 0 < H < 1.
\end{cases}
\tag{B6}
$$

Hence, as $K \searrow 0$, the relative noise decreases to zero linearly if $H > 2$, sublinearly if $1 < H < 2$, or tends to a positive constant $1 - H$ if $0 < H < 1$. High cooperativity in feedback in decay rate thus improves its performance in the strong-feedback regime. Even in the worst case scenario $0 < H < 1$ in terms of noise control, the limiting value $1 - H$ of relative noise is less than the limiting value $1/(1 + H)$ of the small-noise prediction (A5) for the relative noise.

References

1. Abramowitz, M., Stegun, I.: Handbook of Mathematical Functions with Formulas, Graphs, and Mathematical Tables. National Bureau of Standards, Washington, D.C. (1972)
2. Becskei, A., Serrano, L.: Engineering stability in gene networks by autoregulation. Nature **405**, 590–593 (2000)
3. Bokes, P., King, J., Wood, A., Loose, M.: Multiscale stochastic modelling of gene expression. J. Math. Biol. **65**, 493–520 (2012). https://doi.org/10.1007/s00285-011-0468-7
4. Bokes, P., King, J., Wood, A., Loose, M.: Transcriptional bursting diversifies the behaviour of a toggle switch: hybrid simulation of stochastic gene expression. Bull. Math. Biol. **75**, 351–371 (2013)
5. Bokes, P., Lin, Y., Singh, A.: High cooperativity in negative feedback can amplify noisy gene expression. Bull. Math. Biol. **80**, 1871–1899 (2018). https://doi.org/10.1007/s11538-018-0438-y
6. Bokes, P., Hojcka, M., Singh, A.: Buffering gene expression noise by MicroRNA based feedforward regulation. In: Češka, M., Šafránek, D. (eds.) CMSB 2018. LNCS, vol. 11095, pp. 129–145. Springer, Cham (2018). https://doi.org/10.1007/978-3-319-99429-1_8

7. Bokes, P., King, J.R., Wood, A.T., Loose, M.: Exact and approximate distributions of protein and mRNA levels in the low-copy regime of gene expression. J. Math. Biol. **64**, 829–854 (2012)
8. Bokes, P., Singh, A.: Gene expression noise is affected differentially by feedback in burst frequency and burst size. J. Math. Biol. **74**, 1483–1509 (2017)
9. Bortolussi, L.: Hybrid behaviour of Markov population models. Inf. Comput. **247**, 37–86 (2016)
10. Cai, L., Friedman, N., Xie, X.: Stochastic protein expression in individual cells at the single molecule level. Nature **440**, 358–362 (2006)
11. Cardelli, L., Kwiatkowska, M., Laurenti, L.: A stochastic hybrid approximation for chemical kinetics based on the linear noise approximation. In: Bartocci, E., Lio, P., Paoletti, N. (eds.) CMSB 2016. LNCS, vol. 9859, pp. 147–167. Springer, Cham (2016). https://doi.org/10.1007/978-3-319-45177-0_10
12. Cinquemani, E., Milias-Argeitis, A., Summers, S., Lygeros, J.: Local identification of piecewise deterministic models of genetic networks. In: Majumdar, R., Tabuada, P. (eds.) HSCC 2009. LNCS, vol. 5469, pp. 105–119. Springer, Heidelberg (2009). https://doi.org/10.1007/978-3-642-00602-9_8
13. Dar, R.D., et al.: Transcriptional burst frequency and burst size are equally modulated across the human genome. Proc. Natl. Acad. Sci. U.S.A. **109**, 17454–17459 (2012)
14. Dessalles, R., Fromion, V., Robert, P.: A stochastic analysis of autoregulation of gene expression. J. Math. Biol. **75**, 1253–1283 (2017). https://doi.org/10.1007/s00285-017-1116-7
15. Friedman, N., Cai, L., Xie, X.: Linking stochastic dynamics to population distribution: an analytical framework of gene expression. Phys. Rev. Lett. **97**, 168302 (2006)
16. Hernandez, M.A., Patel, B., Hey, F., Giblett, S., Davis, H., Pritchard, C.: Regulation of BRAF protein stability by a negative feedback loop involving the MEK-ERK pathway but not the FBXW7 tumour suppressor. Cell. Signal. **28**, 561–571 (2016)
17. Hinch, E.J.: Perturbation Methods. Cambridge University Press, Cambridge (1991)
18. Kumar, N., Platini, T., Kulkarni, R.V.: Exact distributions for stochastic gene expression models with bursting and feedback. Phys. Rev. Lett. **113**, 268105 (2014)
19. Kurasov, P., Lück, A., Mugnolo, D., Wolf, V.: Stochastic hybrid models of gene regulatory networks – a PDE approach. Math. Biosci. **305**, 170–177 (2018)
20. Lin, G., Yu, J., Zhou, Z., Sun, Q., Jiao, F.: Fluctuations of mRNA distributions in multiple pathway activated transcription. Discrete Contin. Dyn. Syst.-B (2018). https://doi.org/10.3934/dcdsb.2018219
21. Lin, Y.T., Doering, C.R.: Gene expression dynamics with stochastic bursts: construction and exact results for a coarse-grained model. Phys. Rev. E **93**, 022409 (2016)
22. McAdams, H., Arkin, A.: Stochastic mechanisms in gene expression. Proc. Natl. Acad. Sci. U.S.A. **94**, 814–819 (1997)
23. Ocone, A., Millar, A.J., Sanguinetti, G.: Hybrid regulatory models: a statistically tractable approach to model regulatory network dynamics. Bioinformatics **29**, 910–916 (2013)
24. Pájaro, M., Alonso, A.A., Otero-Muras, I., Vázquez, C.: Stochastic modeling and numerical simulation of gene regulatory networks with protein bursting. J. Theor. Biol. **421**, 51–70 (2017)
25. Singh, A., Hespanha, J.P.: Optimal feedback strength for noise suppression in autoregulatory gene networks. Biophys. J. **96**, 4013–4023 (2009)

26. Singh, A., Hespanha, J.P.: Reducing noise through translational control in an auto-regulatory gene network. In: 2009 American Control Conference, ACC 2009, pp. 1712–1717. IEEE (2009)
27. Sundqvist, A., Ericsson, J.: Transcription-dependent degradation controls the stability of the SREBP family of transcription factors. Proc. Natl. Acad. Sci. U.S.A. **100**, 13833–13838 (2003)
28. Thattai, M., van Oudenaarden, A.: Intrinsic noise in gene regulatory networks. Proc. Natl. Acad. Sci. U.S.A. **98**, 151588598 (2001)

Extracting Landscape Features from Single Particle Trajectories

Ádám M. Halász[1($\boxtimes$)], Brandon L. Clark[1], Ouri Maler[1,2],
and Jeremy S. Edwards[3,4]

[1] Department of Mathematics, West Virginia University, Morgantown, WV, USA
halasz@math.wvu.edu
[2] VERIMAG, Université Grenoble Alpes, Grenoble, France
[3] N. M. Center for the SpatioTemporal Modeling of Cell Signaling,
University of New Mexico, Albuquerque, NM, USA
[4] Departments of Chemical and Biological Engineering, Chemistry and Chemical
Biology, Molecular Genetics and Microbiology, University of New Mexico,
Albuquerque, NM, USA

Abstract. The predictive power of dynamical models of cell signaling is often limited due to the difficulty in estimating the relevant kinetic parameters. Super-resolution microscopy techniques can provide in vivo trajectories of *individual* receptors, and serve as a direct source of quantitative information on molecular processes. Single particle tracking (SPT) has been used to extract reaction kinetic parameters such as dimer lifetimes and diffusion rates. However, signaling models aim to characterize kinetics relevant to the entire cell while SPT follows individual molecules in a small fraction of the cell. The gap in resolution can be bridged with spatial simulations of molecular movement, validated at SPT resolution, which are used to infer effective kinetics on larger spatial scales.

Our focus is on processes that involve receptors bound to the cell membrane. Extrapolating kinetics observed at SPT resolution must take into account the spatial structures that interferes with the free movement of molecules of interest. This is reflected in patterns of movement that deviate from standard Brownian motion. Ideally, simulations at SPT resolution should reproduce observed movement patterns, which reflect the properties and transformation of the molecules as well as those of the underlying cell membrane.

We first sought to identify general signatures of the underlying membrane landscape in jump size distributions extracted from SPT data. We found that Brownian motion simulations in the presence of a pattern of obstacles could provide a good qualitative match. The next step is to infer the underlying landscape structures. We discuss our method used to identify such structures from long single particle trajectories that are obtained at low density. Our approach is based on deviations from ideal Brownian motion and identifies likely regions that trap receptors. We discuss the details of the method in its current form and outline a framework aimed at refinement using simulated motion in a known landscape.

Keywords: Membrane receptors · EGF · Brownian motion

M. Češka and N. Paoletti (Eds.): HSB 2019, LNBI 11705, pp. 98–116, 2019.
https://doi.org/10.1007/978-3-030-28042-0_7

1   Introduction

Cell signaling plays an important role in the normal functioning of cells and in health conditions such as cancers and immune diseases [1,5,12]. Cell signaling models, capturing the transformations of many different molecular species, are one of the most successful applications of dynamical systems biology [3,10,19,21]. The predicting power of these highly nonlinear dynamical models is often limited due to difficulties in estimating the relevant kinetic parameters. It is difficult to infer the dynamics of individual processes in a living system by separating them from the rest of system. Due to many factors, including the spatial structure of the cell, parameters that appropriately characterize a given process within a living cell can be very different from those observed when reproducing the process in synthetic ("in vitro") conditions. Advances in molecular resolution imaging could provide a way around this, by estimating reaction parameters *in vivo*, from the movement of individual molecules. This is a realistic possibility for *membrane-bound receptors* important for cancers and other major health problems [12,20,23,34] whose two dimensional movement can be reliably mapped using currently available technology [2,9,25].

Super-resolution microscopy provides in vivo trajectories of receptors and other bio-molecules labelled with fluorescent tags [22,25]. The behavior of individual molecules in a small fraction of the cell is "too detailed" – well beyond the spatial resolution relevant to current quantitative models of cell signaling [10,21,28]. Single particle tracking (SPT) is used to infer diffusion coefficients and dimerization rates [9,25,26,32]. However, the characteristics of movement as well as the identification of dimerization/dissociation events observed at SPT resolution are complicated by transient trapping of molecules in small domains [11,22,26].

Here we investigate the possible signatures and present a method to identify such structures, based on deviations from ideal Brownian motion and outline a framework aimed at refinement using simulated motion in a known landscape. Our first goal is to understand and identify the non-Brownian ("anomalous") aspects of particle movement. The study of jump size distributions from Brownian motion simulations in artificial landscapes of barriers led us to identify the "hockey-stick" shaped square displacement distributions (Fig. 1B) as a likely signature of the presence of trapping domains. We found that the qualitative features were consistently reproduced, but the quantitative aspects of the distributions were sensitive to the detailed properties of the domains. The realization of the specificity of the jump size distributions led us to the second aim, of identifying an underlying landscape of (likely) trapping domains from specific trajectories. The resulting method consists of a scoring process that identifies locations where particle movement is slow, and a domain reconstruction part that results in geometric footprints of the domains.

The paper is organized as follows. In the remainder of this section we provide some background on dynamical modeling of cellular processes, describe the relevant experimental methods in the field, and define the problem of interest. Section 2 contains a summary of basic mathematical aspects of diffusion and Brownian motion, along with a basic observations and hypotheses regarding the quasi-Brownian behavior of membrane bound receptors a observed by singe plarticle tracking. Section 3 discusses simulation methods used to model signal initiation by membrane receptors and to analyze jump size distributions extracted from particle tracking. This section ends with the description of the domain reconstruction method we developed and used in recent work. Section 4 begins with the discussion of comparing jump size distributions from simulated movement in landscapes with semipermeable barriers and actual SPT data. We present an example of a set of reconstructed domains from trajectories, similar to what we did in several projects where these reconstructed domains were used in model simulations. Finally, we outline the simulation environment we constructed aimed at generating synthetic trajectories for the refinement of the domain identification process, and ideas for future applications.

1.1   Background and Motivation

Dynamical models of cell signaling aim to capture the molecular processes that take place in a cell, from the appearance of a stimulus to the triggering of the cell's response. The resulting dynamical systems are typically high dimensional [4,6,15,19]. In spite of progress [7,36,37] toward the identification of robust behaviors, there are many possible steady states and regimes, and the emerging behaviors are sensitive to parameter values. For signaling models, this is further compounded by the need to capture the behavior of the entire cell (or at least the parts involved from the stimulus to the implementation of the cellular response). Fully stochastic and/or spatially resolved models are impractical, and a useful signaling model will rely on effective kinetic constants that reflect an *average* behavior.

Biochemical experimentation tends to focus on identifying and establishing the role of a given substance. This is often achieved *in vitro* or in organisms specifically modified to enhance and isolate the process of interest. Such setups will often result in conditions that are very different from the *in vivo* context relevant to the signaling process. Values of rate constants found in the literature for the same process might vary by orders of magnitude. It is not realistic to expect large amounts of laboratory resources dedicated to parameterizing dynamical models by repeating almost identical experiments under slightly changed conditions. The perceived benefit is not exciting the way finding a new transcription factor would be.

Interest in systems biology is fueled by advances in genetic technology as well as other technologies that allow molecular level intervention into the functioning of cells. Fluorescent labeling of individual molecular structures involved in signaling, combined with microscopy and automated image processing, provides modalities to interrogate individual molecular processes, *in vivo*, in minimally modified cells. *Flow cytometry* (FC) allows the estimation of the amount (often the number) of tagged molecules in individual cells. Flow cytometry can provide whole cell measurements of individual molecular species, together with their distribution over a population. The time resolution is somewhat limited by the experimental setup, but this modality provides insight into the overall functioning of a signaling pathway or of its elements. *Super-resolution microscopy* can localize fluorophores with a precision of tens of nanometers ($1nm = 10^{-3}\mu m = 10^{-9}m$) and generate trajectories with time resolution from ≈ 50 frames per second [2,22,25,29]. The resulting single particle tracks (SPT) represent a much higher level of detail compared to FC. However, this is often necessary in order to disentangle the steps involved in signaling. It can also potentially connect to molecular dynamics, another level of modeling study concerned with the functionality of bio-molecules as it results from their atomic and spatial structure.

SPT has been previously used to estimate dimerization and dissociation constants and to study transient confinement of diffusing receptors [25,26,32]. At this level of resolution, nano-scale details of spatial organization become important. In particular, the presence of obstacles to the movement of membrane bound receptors result in a movement pattern of *anomalous diffusion*; the particles move consistent with Brownian motion, but the observed diffusion coefficient is dependent on the time scale of observation. Closer examination of the trajectories reveals transient trapping consistent with the presence of obstacles.

We report on our ongoing effort aimed at understanding and modeling the phenomena of transient trapping and anomalous diffusion in SPT. A properly validated microscopic model of diffusion behavior can be used to predict diffusion behavior at longer time and spatial scales. An equivalent diffusion-reaction model can be the basis of abstractions and approximations, leading to simple spatially averaged reaction models that can be used for cell signaling.

1.2   Statement of the Problem and Related Work

We have a set of trajectories resulting from one or more single particle tracking (SPT) experiments. Trajectories represent the consecutive positions of a single molecular entity (such as a receptor, or ligand) throughout its movement and interactions/transformations. Our goal is to infer factors that influence the observed movement, in particular: (i) obstacles and (2) changes in mobility due to interaction with other molecular entities.

An experiment consists of a recording of a *sample*, and yields a set of trajectories. Images (frames) are recorded at fixed time intervals. The positions of (point) sources of emitted light are identified during image processing. These are fluorescent tags, which are bound to molecules of a certain type. A given tag may not be detected in every frame of a recording. One trajectory consists of a sequence of positions that have been identified as representing the same fluorescent tag. In addition to the issue of missing points, trajectories are affected by a position uncertainty and the fact that two tags of the same color that are close can not be distinguished. In the absence of interaction with obstacles, we assume that molecules perform Brownian motion in two dimensions, consistent with a "free-space" diffusion coefficient[1]. Obstacles are one dimensional barriers that either completely block movement or allow crossing with a certain probability $p_{cross} < 1$, which may be different for the two directions. Additionally, tagged molecules may join larger aggregates, resulting in reduced mobility.

The main mathematical problem is to identify a set of obstacles that would result in the observed trajectories, assuming that the intrinsic diffusion coefficient of a particle does not change.

Related Work: There are two directions of investigation we are aware of. First, the methodology of extracting trajectories from single fluorescent proteins [16, 30,35] and the related methods developed for identifying confinement (trapping) of individual particles [26,33] rely on comparing jump sizes with the normal distributions expected from Brownian motion. In particular, co-confinement was identified in [26] using a hidden Markov model that relied on the mutual distance between two particles.

Another direction [8,27] relies on *high-density* single molecule imaging, which results in shorter trajectories (due to the difficulty of identifying consecutive positions of the same particle). Due to the different experimental approach this results in a higher resolution mapping of the mmbrane landscape but the shorter trajectories do not allow simultaneous investigation of binding and unbinding events and the extraction of dimerization/dissociation rates.

2   Preliminaries

2.1   Brownian Motion/Ideal Diffusion

Brownian motion (BM), taken in the mathematical sense, is a type of random process. In two spatial dimensions, the components of the position vector $r(t) = (x(t), y(t))$ representing a moving point [-like object], are continuous random variables, and the components $\Delta x, \Delta y$ of the displacement vector $\Delta r = r(t + \Delta t) - r(t)$ over any interval $(t, t + \Delta t)$ are random variables distributed according to the PDF

$$f(\Delta x, \Delta y; \Delta t) = \frac{1}{4\pi D \Delta t} \exp\left(-\frac{\Delta x^2 + \Delta y^2}{4D\Delta t}\right). \tag{1}$$

The x- and y-displacements are independent and each follows a normal distribution with variance $\sigma^2 = 2D\Delta t$. The parameter D is a measure of the mobility of the particle. A diffusing substance can be represented as a set of n particles that moves consistent with (1). We can describe a set of n Brownian particles by the *sum* of the individual localization probability density functions that each

---

[1] Brownian motion, as well as confinement to the cell membrane, result from interaction with much smaller molecules such as the lipids that form the membrane. We do not explicitly address this level of interaction.

follow (1). This joint localization density function $\rho(x, y; t)$ verifies a diffusion equation:

$$\frac{\partial \rho}{\partial t} = D \left(\frac{\partial^2 \rho}{\partial x^2} + \frac{\partial^2 \rho}{\partial y^2} \right). \tag{2}$$

Brownian Motion is central to any discussion of random movement. It is a consequence of the Central Limit Theorem [17] that *any isotropic random walk approaches ideal Brownian motion* in the limit of large time (or spatial) scales. In a random walk, the position of the moving object is updated in discrete steps, at finite time intervals. If the displacement along one direction, over a characteristic time τ, has mean zero and standard deviation σ_0, then the distribution of displacements over time T approaches a normal with standard deviation $\sigma(T) = \sqrt{2T \cdot D_{\text{eff}}}$ as $T \gg \tau$ where D_{eff} is given by

$$D_{\text{eff}} = \frac{\sigma_0^2}{2\tau}. \tag{3}$$

If the movement is in two dimensions and the displacements along both x and y verify the assumptions (zero mean and standard deviation $\langle (\Delta X)^2 \rangle = \langle (\Delta Y)^2 \rangle = \sigma_0/\tau$, then, in the limit $T/\tau \to \infty$, the system approaches ideal Brownian motion with diffusion coefficient given by Einstein's relation (3).

However, perfect Brownian motion is a mathematical idealization. In summary, we could say that all *physical* random motion looks Brownian if examined on a long enough time scale. On the other hand, the model becomes unphysical over short time scales. Thus, examination over a short enough time scale should reveal that any apparently Brownian motion is in reality, "anomalous".

Jump Size Distributions. When analyzing trajectories, one often compares the distribution of displacements $\Delta r = r(t + \tau) - r(t)$ over some observation time τ with the Brownian PDF (1). The most commonly used observable is the *square displacement* $s \equiv (\Delta r)^2 = |\Delta r|^2 = (\Delta x)^2 + (\Delta y)^2$, whose distribution for fixed observation time τ is exponential[2]:

$$f_{\text{SD}}(s) = \frac{1}{4D\tau} \exp \left(-\frac{s}{4D\tau} \right) \tag{4}$$

The *mean square displacement* (MSD) is the expectation of (4) and is consistent with the standard deviation of the two components $\sigma^2 \equiv 2D\tau = \langle (\Delta x)^2 \rangle = \langle (\Delta y)^2 \rangle$:

$$\langle s \rangle = \langle (\Delta r)^2 \rangle = 4D\tau = 2\sigma^2. \tag{5}$$

Equation (5) provides a simple check of the Brownian nature of trajectories. We will also use the distribution of square displacements (SDD) for a given τ.

---

[2] not to be confused with the distribution of the magnitude of the displacement $f(\Delta r) \propto (\Delta r) \exp(-(\Delta r)/4D\tau)$.

2.2   Diffusion and Obstacles

The movement of membrane bound proteins is similar to that of particles suspended in a liquid solution; the diffusive behavior is the result of interaction with the much smaller lipid molecules that form the membrane. These effects are not individually discernible on the scale of tens of nanometers that is relevant to our discussion, and therefore we assume that, in the absence of obstacles[3] the movement of membrane bound receptors can be characterized as Brownian with a "free space" diffusion coefficient D_0.

The experimental picture, derived from single particle tracking, indicates deviations from (5). The mean square displacement (as shown in Fig. 1A) initially increases linearly with the observation time T_{obs}, but the slope is reduced as T_{obs} increases. The corresponding SDD (Fig. 1B) are consistent with a superposition of two or more distributions, that are somewhat consistent with exponentials corresponding to (4) with different diffusion coefficients.

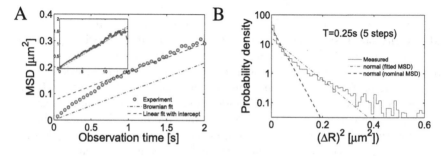

Fig. 1. Square displacement statistics from 1685 SPT trajectories from 21 recordings. The mean square displacement (MSD) as well as the square displacement distributions (SDD) deviate from ideal Brownian motion.

The "anomalous" nature of movement in single particle tracking has been noticed early on [22]. A frequently noticed feature of the trajectories is that particles tend to have intervals of movement that is localized to a relatively small area, interspersed with longer jumps or sequences. There has been a vigorous debate in the biology/biophysics literature on the origin of this phenomenon. Elongated protein structures of varying thickness, called actin filaments, that run along the membrane or are part of the cytoskeleton (a rigid support structure that ensures the shape of the cell, stretching the soft and flexible membrane similarly to a large tent) likely act as physical barriers that impede the movement of membrane bound proteins such as the receptors. The resulting grid of barriers creates a partition of the cell membrane, much like fences that divide grazing land into corrals.

---

[3] ... of comparable size, 10 nm.

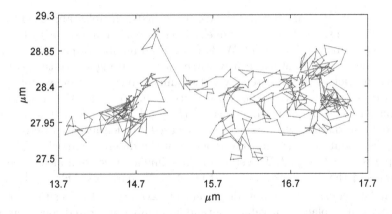

Fig. 2. Two sample SPT trajectories exhibiting typical features of anomalous diffusion consistent with the presence of obstacles and/or trapping. The receptors tend to alternate between confinement to small areas and movement on longer distances.

Corrals would easily explain the observed confinement episodes, but it is hard to see how they may account for the phenomenon of receptor clustering, where receptors of one type tend to accumulate in small areas, consistent with the confinement zones seen in SPT experiments. Lipid rafts are membrane patches whose physico-chemical composition differs from the surroundings, due to the shape and electric charge distribution of the respective lipid molecules. Similar interactions might result in an affinity between specific types of receptors and lipid rafts. The latter would then act as transient traps for receptors, which could move in and out of these attractive domains, with a probabilistic exit penalty.

Obviously, these two possible explanations are not mutually exclusive and are by no means the only ones. Another important factor in modulating the mobility of membrane bound proteins is their participation in aggregates, less stable structures consisting of several molecular entities, receptors, their ligands, and associated proteins that are part of the orchestration of the signaling process. Thus, a change in the mobility of a labelled receptor may reflect trapping, interaction with other proteins, or its association/exit from a (larger, therefore less mobile) aggregate.

3   Methods and Models

3.1   Simulation

The spatial simulation approach developed in our recent work ([31,32], and others) relies on a number of standard methods, self-implemented in programs developed for the purpose of the specific application.

Brownian Motion (BM) Simulator. We use a first-principles approach [31], in that our motion simulation algorithm emulates Brownian motion using the exact expression for displacements (1). We follow a fixed number of particles, N_p, and generate their positions at consecutive times $t_0, t_1, \ldots$ using a fixed time step, Δt.

Each particle is represented by a two dimensional position vector, which evolves over time reflecting the movements of the particle. For instance, the position vector of the k-th particle is $\boldsymbol{r}^{(k)}(t) = (x^{(k)}(t), y^{(k)}(t))$. It is helpful to point out the distinction between integers used to identify the discrete times (when the updates occur) and the labeling of the different particles. The time at step j is $t_j = t_0 + j\Delta t$. The state of the (simulated) system at a given time t_j consists of the XY coordinates of the set of N_p particles, $\left\{ \boldsymbol{r}^{(k)}(t_j) \right\}_{k=1\cdots N_p}$. Each update consists of adding displacements $\Delta x, \Delta y$ to the positions of each particle. Each displacement value is a random number generated consistent with the PDF (1). At every update, the system time is advanced by the amount of one step; for each of the N_P particles, the X and Y coordinates are by amounts Δx and Δy respectively, generated as described above.

Reactions. For a more complete discussion of stochastic simulation of chemical reactions, we refer the interested reader to textbooks and reviews or to [14]. We use a stochastic, agent- and rule-based approach in our fully detailed simulations.

In an *agent-based* simulation, chemical species is a discrete label that characterizes a given particle (agent). A particle may change its species without interacting with others (similar to radioactive decay). More complicated situations arise due to oligomerization. Agents maintain their identity even when they become bound in such a group.

Rule-based modeling. The number of species and reactions induced by the combinations of proteins participating in a signaling network, is potentially enormous, a phenomenon known as combinatorial complexity. The traditional method of specifying the kinetics for a chemical reaction system requires a list of all possible species and reactions that can potentially occur in the system. The alternative is a rules-based approach [24] where types of transformations are specified, and reactions, and species, are generated accordingly. Combinations of agents bound to each other in various internal states correspond to species in the traditional sense.

3.2   Analysis of Jump Size Distributions

As illustrated in Fig. 1B, distributions of square displacements (SDD) derived from experimental tracks deviate from the simple exponential (4). The "hockey stick" shaped distribution can be interpreted as a superposition of two exponentials,

$$f_{\mathrm{SD}}(s) = \frac{A_1}{4D_1\tau} \exp\left(-\frac{s}{4D_1\tau}\right) + \frac{A_2}{4D_2\tau} \exp\left(-\frac{s}{4D_2\tau}\right); \quad D_1 > D_2. \quad (6)$$

The steeper negative slope seen on the semi-logarithmic scale ($1/4D_2\tau$) corresponds to *slower* motion, labeled "nominal MSD") and dominates the shorter displacements. The component corresponding to the higher diffusion coefficient, hence lower slope, dominates the longer displacements.

Looking at the mean square displacements (MSD, Fig. 1A), the obvious feature is that *shorter* time scales correspond to a higher slope. Here the Brownian model would predict $\langle(\Delta R)^2\rangle = 2DT_{\text{obs}}$, therefore the initial higher slope corresponds to *faster* movement for shorter time scales.

Deconvolution. One avenue for a deeper analysis of the SDD consists of fitting a number of exponential distributions. One challenge in dealing with data that is roughly exponentially distributed is that building histograms to compare with a theoretical PDF involves difference of orders of magnitude in terms of the expected counts. It is intuitively helpful to look at the distribution pertaining to the logarithms of the observed square displacements; if the quantity s follows the PDF $f(s) = 1/\mu \exp(-s/\mu)$, then its logarithm $\lambda = ln(s)$ will follow the PDF $g(\lambda) = (e^\lambda/\mu)e^{-e^\lambda/\mu}$. This function approaches zero asymptotically at $\lambda \to \pm\infty$ and has a single maximum at $\lambda = ln(\mu)$ which corresponds to $s = ln(\lambda) = \mu$.

3.3   Domain Reconstruction Algorithm

The domain reconstruction algorithm (DRA) we developed uses a set of trajectories from a given experiment to identify areas in the field of view where the movement of particles is slower. It relies exclusively on the distribution of jump sizes (displacements) between points in the same trajectory and can be summarized as follows.

1. Identify "slow" points
   (a) Build distributions of displacements over $m_1 < m_2 \cdots m_p$ frame intervals
   (b) For each point $r_k^{(j)}$ in each trajectory j:
       i. Find the percentage rank $\rho_i(r_k^{(j)})$ of displacements over $\{m_i\}_{i=1\cdots p}$ frame intervals.
       ii. Construct a weighted score $w(r_k^{(j)}) = \sum_{i=1}^{p} \alpha_i \rho_i(r_k^{(j)})$
   (c) Obtain the distribution of weighted scores over the entire sample, and choose a cutoff to define points associated with slower movement
2. Group the slow points into distance based clusters
   (a) Compute the mutual distances between all pairs of slow points
   (b) Two points A, B are "connected" if
       i. $|r_A - r_B| < L$ (their distance is less than L)
       ii. There is another point C that is connected to both A and B[4]
3. Build a (non-convex) envelope around the clusters consistent with L
   (a) *For each cluster*, build a graph with edges connecting pairs of points whose distance is $< L$

---

[4] This implies that two points will be connected if there is a connecting path $\{A = M_0, M_1, \cdots M_q = B\}$ through any number of other points and no edge longer than L, $|r_{M_j} - r_{M_{j+1}}| < L$.

(b) identify the outer edges of the graph and form a closed poly-line[5]
  i. Choose the outer-most point in some direction u, let that be A_0 (r_0)
  ii. From the points in the cluster that are connected to A_0, identify the point A_1 for which the angle $A_1 A_0 u$, measured counter-clockwise from u is the smallest. This will be the next point in the inner contour.
  iii. Choose the point connected to A_2 such that the angle $A_2 A_1 A_0$ measured counter-clockwise from A_0 is minimal.
  iv. Repeat the previous step, each time adding a new point to the inner contour.
  v. Stop when the initial point A_0 is reached.
(c) Build a padding around the outer edges
  i. Using each outer edge, construct a rectangle outside the graph, whose other two sides have length $L/2$.
  ii. If the angle between two adjacent outer edges is $<180°$, clip the corresponding rectangle sides at their intersection
  iii. If the angle is $>180°$, connect the outer rectangle sides with arc of circle centered on the common vertex

Motivation. Assuming that at least some of the variations in mobility result from the underlying landscape, we are interested in finding correlations between the mobility of receptors, as reflected by their jump sizes and spatial locations. The distance between point r_k in a trajectory, corresponding to frame k, and the points r_{k-1}, r_{k+1} immediately before and after it carry some information on the landscape at r_k. Since these jump sizes are random variables, a short jump does not necessarily imply the presence of an obstacle, but makes it more likely that an obstacle is present in the vicinity of r_k.

The time between consecutive frames is set by the experiment. One can make the same argument for jumps over several frames, $\Delta r_{km} = r_k - r_{k+m}$. If confining domains or other obstacles are present, then the relevant time scale τ^* should be such that the mean displacement $2\sqrt{D\tau^*} \approx \ell$, where ℓ is the linear size of the domains or some typical distance between obstacles. For time intervals of length τ smaller than τ^*, the likelihood that a given step involves hitting an obstacle decreases with τ/τ^*. On the other hand, jumps over $\tau \gg \tau^*$ will almost certainly involve obstacles, so we won't be able to distinguish any locations.

Scoring. Since the length scale ℓ is not a priori known, we constructed a combined score for points in trajectories as follows. Assume a set of trajectories $T_1, \cdots T_{N_{traj}}$ from the same experiment or group of experiments, collected at frame interval τ. We choose a set of p integers $m_1 < m_2 < \cdots m_p$, and collect jump size distributions from each trajectory $T_j = \{r_1^{(j)}, r_2^{(j)}, \ldots r_{N_{steps}}^{(j)}\}$. The set $\mathcal{D}_m^{(j)}$ of displacements of length $t_{obs} = \tau m$ collected from trajectory T_j will be a *non-overlapping* subset of all possible displacements $\left\{ |\Delta r_{k,k+m}^{(j)}| \right\}_{1 \leq k \leq N_{steps}-m}$.

---

[5] This is not completely unique, but will always provide a closed polygonal line.

Do this for each trajectory to collect the merged and sorted set of displacements of m time steps $\mathcal{D}_m = \mathcal{D}_m^{(1)} \cup \mathcal{D}_m^{(2)} \cdots \mathcal{D}_m^{(N_{traj})}$.

Once the jump sizes have been compiled, the individual points from each trajectory are ranked, based on the jumps of each length m that begin and end on that point; for example, the point $r_k^{(j)}$ gets the average of relative ranks of $|\Delta r_{k,k+m}^{(j)}|$ and $|\Delta r_{k-m,k}^{(j)}|$ in the set $\mathcal{D}_m$. For every point, the relative rank is calculated by comparing the point's jump size to all the other points' jump sizes for a specific step size. The jump sizes for a specific step count m are sorted in order from smallest to largest. The rank of a specific point is where it falls in that order; the relative rank is obtained by dividing by the number of entries in $\mathcal{D}_m$. This is repeated across all points in each trajectory, and for each step size $m_1 < m_2 < \cdots m_p$. Finally, a weighted average is used to determine the overall score for each point

$$\omega(r_k^{(j)}) = \sum_{i=1}^{p} \alpha_i \rho_i(r_k^{(j)}), \tag{7}$$

where $\rho_p(r_k^{(j)})$ stands for the relative rank of the point using m_p time steps. Some points may not have a score for all p step sizes due to the aforementioned holes in the trajectories. In this case we omit the missing jumps from the averaging (7).

The combined score (7) is interpreted as a measure of the likelihood that a point is close to an obstacle. We use it to sort the points in a group of trajectories obtained from the same recording by choosing a cutoff value and identify a set of "slow" points. We found that, by choosing the weights appropriately, the distribution of scores often becomes bimodal, which is helpful in choosing a cutoff.

Clustering and Domains. More often than not, when applying to experimental trajectories, the "slow" points identified as described above do tend to collect in areas preferentially visited by the particles. This is reflected in an uneven distribution, so that the slow points form clusters. We apply a cluster identification algorithm based on hierarchical distance based clustering [13] to partition the slow points into clusters. This algorithm relies on a length parameter L. In practice, this is adjusted on a trial and error basis.

For each identified cluster, we construct a geometric footprint (shape) around the member points to provide us with an estimate of the area and perimeter for the cluster. This footprint can be identified with an underlying physical support. This was described in detail elsewhere ([31] as well as [13]). Briefly, the points in a cluster are used to define the vertices of a graph, whose edges link point whose distane is less than the length parameter L. By the definition of the clusters, this graph must be connected. An outer contour (subset of the graph) is identified by walking around the graph in one direction (e.g. clockwise). This contour is then padded with circles of radius $L/2$ around the vertices and rectangles around the outer edges. The resulting padded contour is exported as a poly-line.

4   Results and Discussion

The efforts discussed here address three objectives, all related to understanding the movement of membrane bound receptors (and other molecules) as observed in low density single particle tracking experiments. These are (i) a plausible explanation for the observed features of anomalous diffusion (ii) an efficient methodology for the identification of confining domains (iii) validation, consistency checks, and optimization of the existing methods.

Plausible Explanation of Anomalous Diffusion: The anomalous features of receptor diffusion consist of deviations of the mean square displacement from (5) and of the square displacement distributions from (4). Using Brownian motion simulations in a synthetic landscape with rectangular shaped trapping domains, we have obtained square displacement distributions with "hockey stick" shapes similar to the experimentally observed ones (Fig. 3). The corresponding MSD curves exhibit sub-diffusion similar to 1A, albeit to a significantly lesser degree than the experimental ones.

These simulation results provide the plausible explanation we sought. Particles get transiently trapped (confined) to relatively small domains, whose diameter is comparable to the mean square displacements over a few frame intervals. Trapped particles will exhibit smaller jumps compared to the free ones. The observed longer jumps correspond to particles that are free (at least for part of the observation time) and are thus moving consistently with the free space diffusion constant; hence the apparent higher diffusion rate observed at longer distances, which gives the hockey stick shape to the square displacement distributions.

Implementation of a Method to Identify Confining Domains: The procedure outlined in Sect. 3.3 was implemented in Matlab and identifies likely confining areas. The efficiency is reasonable in that data corresponding to a typical set of experiments (a few days of recording) is processed in not much more than 1 hour. This implementation has been used to generate simulation landscapes for several projects [18, 31]. Several additional features, such as computations of domain areas, perimeter lengths, geometric form factors, are implemented and can be used to collect statistics over larger groups of recordings.

Many applications and consistency checks are yet to be performed in a consistent fashion. To test and further refine the hypothesis of confining domains, we need to build statistics of domain sizes and form factors; the area corresponding to confining domains and the fraction of the trajectories that are localized in them; co-localization in the same domain of different trajectories or repeated visits by the same trajectory separated by long time intervals. These can then be compared to meaningful null hypothesis of uniform, unimpeded diffusion and several species of particles with different mobilities.

Toward a Framework for Simulation Based Validation: The most obvious open question is that of self-consistency of the domain identification method. Synthetic trajectories from simulations of diffusion in a landscape of domains may be used as input to the domain identification process. This feedback could then be used to refine the methods.

We can report initial results in this direction. The original difficulty in performing this self-consistency check is that our simulations are quite complex – in addition to diffusion in domains, we also include many different chemical interactions, which add greatly to the computational cost. Therefore, we started the separate development of a Brownian simulation in a landscape of random barriers. Results from simulations in this framework are shown in Fig. 6.

4.1 Signature of Confinement or Just Different Mobility States?

The simplest explanation of a two-exponential distribution (6) is that we have two populations of particles with different diffusion coefficients. Assuming that the labeling is specific to one receptor species, we might still see two mobilities if the receptors could form dimers that are less mobile.

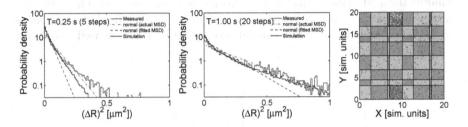

Fig. 3. Simulations with rectangular domains can approximate the observed square displacement distributions.

Another possible explanation is *transient confinement* in domains whose linear size ℓ is within the range of mean displacements $\sqrt{\langle(\Delta R)^2\rangle}$. Particles move freely within these domains, with diffusion coefficient D_0. Over time scales τ where the diffusion distance (square root of the MSD) becomes comparable to the linear size of the domains ($2\sqrt{D_0\tau} \approx \ell$), free movement is impeded. The particles may escape with a certain probability, but a fraction will be unable to move beyond the boundaries. This is reflected in the inflection in the MSD curve: as T_{obs} increases, the fraction of particles that have been turned back at the boundary increases, leading to lower averages. If confinement was perfect, the MSD would approach a horizontal line.

Displacement statistics at fixed time reflect the relative frequency of jumps of each size. Jumps that exceed the domain size ℓ are not impossible, but their likelihood is smaller, because the jump requires an escape event. Once escaped, a particle gains an additional are of free diffusion. Particles that have successfully escaped (during or before the observation interval) and are free to diffuse over distances larger than ℓ, and therefore their square displacements are consistent with higher mobility (hence the flatter slope). By contrast, particles that remain confined over the observation time will exhibit displacement statistics that are bounded. Since the shape and diameter of the confining domains varies, the corresponding jump distribution is not dramatically different from normal.

Simulations with simple landscapes of barriers did produce distributions similar to the experimental ones, as illustrated in Fig. 3. This result supports the obstacle based interpretation, but does not exclude a model with particles having different mobility characteristics, that may or may not change from one state to another. Furthermore, we know that such changes occur in the course of signaling, so we would like to be able to (1) identify mobility changes and (2) clarify whether there is a correlation between the spatial location and movement patterns of particles.

4.2   Reconstructed Domains

Figure 5 illustrates a set of trajectories and the domains inferred from them. A few other trajectories were omitted for clarity. Red dots indicate "slow" points, including those identified from the trajectories that were omitted. The clustering leaves some of these points in pairs or as singletons. We used domain maps derived from the DRA in model simulations for ErbB [18,31] as well as ongoing work on pre B cell receptors.

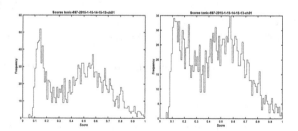

Fig. 4. Two different cumulative scores for the same sample.

The domain reconstruction algorithm (DRA) as described above is able to identify likely areas of high affinity. This algorithm still relies on a number of parameters and choices that are, at this point, chosen on an ad hoc basis. There is no clear guidance for the choice of the frame intervals $m_1, m_2, \cdots$ and especially the of corresponding weights α_i. Figure 4 illustrates the effect of selecting different weights.

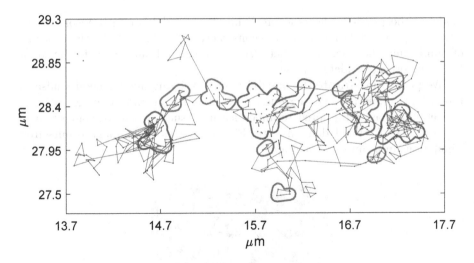

Fig. 5. Illustration of the domain reconstruction algorithm (DRA). The trajectories shown are the same as in Fig. 2. The red/magenta contours are putative attractive domains identified with the DRA. Red dots indicate 'slow' points from the two trajectories as well as from other trajectories in the same recording. (Color figure online)

4.3   Simulation - Analysis - Improvement Feedback

As we outlined in the introduction, the longer term objective/vision/hope is that we can leverage the emerging technology of molecular resolution imaging to construct more realistic, better parameterized, and specifically tuned dynamical models of cell signaling. The aim of the effort discussed in this contribution is to use spatially resolved simulations in conjunction with SPT and other microscopic data to emulate and thus help to understand the processes that the observed particles participate in.

Comparing displacement distributions (SDD mostly) from simulations of two simple obstacle/domain configurations to SPT results, we found that a landscape studded with attractive domains produced a better (still qualitative) match with a data set of FCε receptor trajectories than a landscape with a complete partition into corrals. However, these early results also exhibited a strong dependence on the size and shape of the trapping domains used in the simulations. This prompted us to develop the domain reconstruction algorithm reported in Sect. 3.3. We used the attractive domain contours derived with this method in model simulations for receptors from the ErbB family as well as pre-B cell receptors.

As the reader can hopefully appreciate, the domain reconstruction approach presents many exciting avenues of refinement and automation. To pursue this, we will use spatial simulations of the type developed for our recent work ([31, 32], and others) to compare the input simulation landscape with the structure reconstructed from simulation runs. Logically this is equivalent to regarding the (agent- and rule based spatial stochastic) simulation as a "ground truth" model of the physical phenomenon. While that model is clearly incomplete, a closed loop of simulation-analysis-reconstruction-simulation with the objective of minimizing the discrepancy between the input and output landscape would be an important validation point.

We plan to investigate the possibility of identifying more general landscape types. This raises the exciting question of how to generate a configuration of barriers that has certain predetermined features, such as density, connectedness or lack thereof, closed domains, etc. Figure 6 shows preliminary results from simulations in a random landscape.

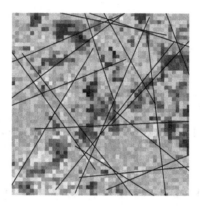

Fig. 6. Localization density heatmap from Brownian motion simulations with random barriers.

References

1. Alberts, B., Bray, D., Lewis, J., Raff, M., Robert, K., Watson, J.D.: Molecular Biology of the Cell, 5th edn. Garland Publishing Inc. (2007)
2. Andrews, N.L., et al.: Actin restricts FcεRI diffusion and facilitates antigen-induced receptors immobilization. Nat. Cell Biol. **10**(8), 955–963 (2008). https://doi.org/10.1038/ncb1755
3. Azuaje, F.: Computational models for predicting drug responses in cancer research. Brief. Bioinf. **18**(5), 820–829 (2017)
4. Barua, D., Hlavacek, W.S., Lipniacki, T.: A computational model for early events in b cell antigen receptor signaling: analysis of the roles of Lyn and Fyn. J. Immunol. **189**(2), 646–658 (2012)
5. Casalini, P., Iorio, M.V., Galmozzi, E., Menard, S.: Role of HER receptors family in development and differentiation. J. Cell. Physiol. **200**, 343–350 (2004)
6. Chen, K.C., Csikasz-Nagy, A., Gyorffy, B., Val, J., Novák, B., Tyson, J.J.: Kinetic analysis of a molecular model of the budding yeast cell cycle. Mol. Biol. Cell **11**(1), 369–391 (2000)
7. Craciun, G., Tang, Y., Feinberg, M.: Understanding bistability in complex enzyme-driven reaction networks. Proc. Nat. Acad. Sci. U.S.A. **103**, 8697–8702 (2006)
8. El Beheiry, M., et al.: A primer on the bayesian approach to high-density single-molecule trajectories analysis. Biophys. J. **110**, 1209–1215 (2016)

9. Erasmus, F.M., et al.: Dynamic pre-BCR homodimers fine-tune autonomous survival signals in B cell precursor acute lymphoblastic leukemia. Sci. Signal. **9**, ra116 (2016)
10. Erickson, K.E., Rukhlenko, O.S., Posner, R.G., Hlavacek, W.S., Kholodenko, B.N.: New insights into RAS biology reinvigorate interest in mathematical modeling of RAS signaling. Semin. Cancer Biol. **54**, 162–173 (2019)
11. Espinoza, F.A., et al.: Insights into cell membrane microdomain organization from live cell single particle tracking of the IgE high affinity receptor FcεRI of mast cells. Bull. Math. Biol. **74**, 1857–1911 (2012)
12. Gounni, A.S.: The high-affinity IgE receptor (FcεRI): a critical regulator of airway smooth muscle cells? Am. J. Physiol. Lung Cell. Mol. Physiol. **291**, L312–321 (2006)
13. Güven, E., Wester, M.J., Wilson, B.S., Edwards, J.S., Halász, Á.M.: Characterization of the experimentally observed clustering of VEGF receptors. In: Češka, M., Šafránek, D. (eds.) CMSB 2018. LNCS, vol. 11095, pp. 75–92. Springer, Cham (2018). https://doi.org/10.1007/978-3-319-99429-1_5
14. Halász, Á.M., Pryor, M.M.C., Wilson, B.S., Edwards, J.S.: Spatiotemporal modeling of membrane receptors. In: Graw, F., Matthäus, F., Pahle, J. (eds.) Modeling Cellular Systems. CMCS, vol. 11, pp. 1–37. Springer, Cham (2017). https://doi.org/10.1007/978-3-319-45833-5_1
15. Hlavacek, W.S., Faeder, J.R., Blinov, M.L., Posner, R.G., Hucka, M., Fontana, W.: Rules for modeling signal-transduction systems. Sci. Signal. **344**, re6 (2006)
16. Jin, S., Verkman, A.S.: Single particle tracking of complex diffusion in membranes: simulation and detection of barrier, raft, and interaction phenomena. J. Phys. Chem. B **111**(14), 3625–3632 (2007)
17. van Kampen, N.G.: Stochastic Processes in Physics and Chemistry. North-Holland, Amsterdam (1992)
18. Kerketta, R., Halasz, A., Steinkamp, M.P., Wilson, B.S., Edwards, J.: Effect of spatial inhomogeneities on the membrane surface on receptor dimerization and signal initiation. Front. Cell Dev. Biol. **4**, 81 (2016)
19. Kholodenko, B.N., Demin, O.V., Moehren, G., Hoek, J.B.: Quantification of short term signaling by the epidermal growth factor receptor. J. Biol. Chem. **274**(42), 30169–30181 (1999)
20. Kitaura, J., et al.: Evidence that IgE molecules mediate a spectrum of effects on mast cell survival and activation via aggregation of the FcεRI. Proc. Natl. Acad. Sci. U.S.A **100**, 12911–12916 (2003)
21. Kolch, W., Halasz, M., Granovskaya, M., Kholodenko, B.N.: The dynamic control of signal transduction networks in cancer cells. Nat. Rev. Cancer **15**(9), 515–27 (2015)
22. Kusumi, A., et al.: Paradigm shift of the plasma membrane concept from the two dimensional continuum fluid to the partitioned fluid: high-speed single-molecule tracking of membrane molecules. Annu. Rev. Biophys. Biomol. Struct. **34**, 351–378 (2005)
23. Lemmon, M.A., Schlessinger, J.: Cell signaling by receptor tyrosine kinases. Cell **141**, 1117–1134 (2010)
24. Li, H., Cao, Y., Petzold, L.R., Gillespie, D.T.: Algorithms and software for stochastic simulation of biochemical reacting systems. Biotechnol. Prog. **24**, 56–61 (2008)
25. Lidke, D.S., Low-Nam, S.T., Cutler, P.J., Lidke, K.A.: Determining FcεRI diffusional dynamics via single quantum dot tracking. Methods Mol. Biol. **748**, 121–132 (2011)

26. Low-Nam, S.T., et al.: ErbB1 dimerization is promoted by domain co-confinement and stabilized by ligand binding. Nat. Struct. Mol. Biol. **18**, 1244–1249 (2011)
27. Masson, J.B., et al.: Mapping the energy and diffusion landscapes of membrane proteins at the cell surface using high-density single-molecule imaging and bayesian inference: application to the multiscale dynamics of glycine receptors in the neuronal membrane. Biophys. J. **106**, 74–83 (2014)
28. Nguyen, L.K., Kolch, W., Kholodenko, B.N.: When ubiquitination meets phosphorylation: a systems biology perspective of EGFR/MAPK signalling. Cell Commun. Signal. **11**, 52 (2013)
29. Ober, R.J., Ram, S., Ward, E.S.: Localization accuracy in single-molecule microscopy. Biophys. J. **86**(2), 1185 (2004)
30. Pezzarossa, A., Fenz, S., Schmidt, T.: Probing structure and dynamics of the cell membrane with single fluorescent proteins. In: Jung, G. (ed.) Fluorescent Proteins II. Springer Series on Fluorescence (Methods and Applications), vol. 12, pp. 185–212. Springer, Heidelberg (2011). https://doi.org/10.1007/4243_2011_24
31. Pryor, M.M., et al.: Orchestration of ErbB3 signaling through heterointeractions and homointeractions. Mol. Biol. Cell **26**(22), 4109–4123 (2015)
32. Pryor, M.M., Low-Nam, S.T., Halász, A.M., Lidke, D.S., Wilson, B.S., Edwards, J.S.: Dynamic transition states of ErbB1 phosphorylation predicted by spatial-stochastic modeling. Biophys. J. **105**(6), 1533–1543 (2013)
33. Rajani, V., Carrero, G., Golan, D.E., de Vries, G., Cairo, C.W.: Analysis of molecular diffusion by first-passage time variance identifies the size of confinement zones. Biophys. J. **100**(6), 1463–1472 (2011)
34. Schlessinger, J.: Ligand-induced, receptor-mediated dimerization and activation of EGF receptor. Cell **110**, 669–672 (2002)
35. Schmidt, T., Schütz, G.J.: Single-molecule analysis of biomembranes. In: Hinterdorfer, P., Oijen, A. (eds.) Handbook of Single-Molecule Biophysics, pp. 19–42. Springer, New York (2009). https://doi.org/10.1007/978-0-387-76497-9_2
36. Shinar, G., Alon, U., Feinberg, M.: Sensitivity and robustness in chemical reaction networks. SIAM J. Appl. Math. **69**(4), 977–998 (2009)
37. Sontag, E.: Monotone and near-monotone biochemical networks. Syst. Synth. Biol. **1**, 59–87 (2007)

A Hybrid HMM Approach
for the Dynamics of DNA Methylation

Charalampos Kyriakopoulos[1], Pascal Giehr[2], Alexander Lück[1(✉)],
Jörn Walter[2], and Verena Wolf[1]

[1] Department of Computer Science, Saarland University, Saarbrücken, Germany
`alexander.lueck@uni-saarland.de`
[2] Department of Biological Sciences, Saarland University, Saarbrücken, Germany

Abstract. The understanding of mechanisms that control epigenetic changes is an important research area in modern functional biology. Epigenetic modifications such as DNA methylation are in general very stable over many cell divisions. DNA methylation can however be subject to specific and fast changes over a short time scale even in non-dividing (i.e. not-replicating) cells. Such dynamic DNA methylation changes are caused by a combination of active demethylation and de novo methylation processes which have not been investigated in integrated models.

Here we present a hybrid (hidden) Markov model to describe the cycle of methylation and demethylation over (short) time scales. Our hybrid model decribes several molecular events either happening at deterministic points (i.e. describing mechanisms that occur only during cell division) and other events occurring at random time points. We test our model on mouse embryonic stem cells using time-resolved data. We predict methylation changes and estimate the efficiencies of the different modification steps related to DNA methylation and demethylation.

Keywords: DNA methylation · Hidden Markov model · Hybrid stochastic model

1 Introduction

All cells of a multi-cellular organism share the same DNA sequence, yet, depending on location and cell type, display distinct cellular programs as a result of controlled gene expression. Hence, the expression of genes is regulated by epigenetic factors such as DNA methylation. In mammals, the methylation of DNA is restricted to the C5 position of cytosine (C) and mostly appears in a CpG di-nucleotide sequence [6,7]. The palindromic nature of CpG positions provides a symmetry which, after DNA replication, allows the stable inheritance of the methylation "signal". Methylation of C to 5-methyl cytosine (5mC) is catalysed by a certain enzyme family, the DNA methyltransferases (Dnmts) [2,29]. Three

© Springer Nature Switzerland AG 2019
M. Češka and N. Paoletti (Eds.): HSB 2019, LNBI 11705, pp. 117–131, 2019.
https://doi.org/10.1007/978-3-030-28042-0_8

conserved and catalytic active family members are associated with the methylation of DNA. Dnmt1 is mainly responsible for maintenance methylation after DNA replication, i.e. the enzyme mainly reestablishes the methylation pattern on the newly synthesized daughter strand [15] according to the inherited information of the parental DNA strand. The enzymes Dnmt3a and Dnmt3b perform *de novo* methylation, where new methyl groups are added on unmethylated Cs [28]. However, there is evidence that this separation of tasks is not definite and that all Dnmts may carry out all tasks to a certain extent [1].

Once established, 5mC can be further modified by oxidation to 5-hydroxymethyl cytosine (5hmC), which can again be oxidized to 5-formyl cytosine (5fC). Then, 5fC can eventually be converted to 5-carboxy cytosine (5caC). All these processes are carried out by the ten-eleven translocation (Tet) enzymes [16,30]. A considerable level of 5hmC can be found in many cells types and its occurrence has been connected to gene regulation as well as genome wide loss of DNA methylation [12,19]. In contrast, 5fC and 5caC are far less abundant and their particular functions remain more illusive. Nevertheless, studies suggest that both oxidized cytosine variants function as intermediates during enzymatic removal of 5mC from the DNA [12].

In general, DNA methylation can be removed in two ways. First, after DNA replication, the absence or blocking of maintenance methylation will result in a passive DNA methylation loss with each cell division (*passive demethylation*). Second, generated 5fC or 5caC is enzymatically removed from the DNA and subsequently replaced by unmodified cytosine (*active demethylation*, see also Fig. 1) [4,13,14,24].

While DNA replication and the associated maintenance methylation happens only once per cell division cycle, *de novo* methylation and the modification processes of 5mC via oxidation, as well as active demethylation, may happen at arbitrary time points. With purely discrete hidden Markov models (HMM), as used in [1] and [9], it is difficult to describe multiple instances of *de novo* methylation or other modification events, such as hydroxylation, formylation and active demethylation, during one cell division cycle. Here, we introduce a hybrid HMM to describe the dynamics of active demethylation. It distinguishes between events at fixed time points (cell division and maintenance methylation) and events at random time points (*de novo* methylation, oxidations, active demethylation). Overall, the hybrid HMM permits a broader applicability compared to our previously discrete models. Particularly interesting are demethylation events which occur during cell differentiation and early embryonic development. In this case, active demethylation plays a significant role since only a very limited number of cell divisions can be observed. By applying our model to a data set of a single copy gene from mouse embryonic stem cells, we were able to accurately predict the frequency of the observable CpG states and the levels of the hidden states, which correspond to the different modified forms of C. Furthermore, we show how to estimate the enzymatic reaction efficiencies using a maximum likelihood approach.

Compared to previous models for describing DNA methylation, we here propose an approach that takes into account both, the rather fixed and deterministic timing of cell division and the random nature of the enzymatic processes. In this way, we are able to present a model that realistically describes the dynamics of DNA methylation and improves our understanding of active demethylation.

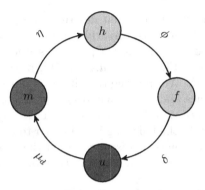

Fig. 1. Schematic representation of *de novo* methylation and the active demethylation loop, where we use the following notation for the methylation states: (Unmethylated) C is denoted by u, 5mC by m, 5hmC by h, and 5fC or 5caC by f. The corresponding enzymatic reaction rates are μ_d (*de novo* methylation), η (oxidation), ϕ (formylation), and δ (demethylation). (Color figure online)

The paper is organized as follows: In Sect. 2 we give the necessary biological and mathematical background, i.e. we explain passive and active demethylation in more detail, describe the model and explain the parameter estimation procedure. In Sect. 3 the results are discussed and in Sect. 4 we conclude our findings.

2   Model

2.1   Passive and Active Demethylation

One can distinguish between two different ways of losing methylation at cytosines: After cell division a new (daughter) strand of the DNA is synthesized. Initially, all cytosines of the daughter strand are unmethylated, while the methylation states of cytosines on the parental strand remain unchanged. Maintenance methylation, which happens during the replication process, is used to re-establish the methylation pattern at the newly synthesized strand. However, the absence or inhibition of maintenance methylation causes a loss of 5mC with each replication step. This DNA replication dependent loss has been termed *passive demethylation*.

For *active demethylation* it is assumed that oxidation of 5mC to 5fC or 5caC via 5hmC and a subsequent enzymatic removal of the oxidative cytosine from the DNA occurs (cf. Fig. 1). Since *de novo* methylation as well as *active demethylation* are replication independent, the loop depicted in Fig. 1 can be traversed multiple times within one cell cycle. In measurements, we see Cs in all stages, i.e., either methylated, oxidized (5hmC/5fC/5caC) or unmodified.

2.2 Hybrid Markov Model

In this section, we present a model that describes the state changes of a single CpG over time. It can be seen as a hybrid extension of previous discrete-time models [1,9]. The (hidden) states of a CpG correspond to the set of all pairs of the four possibilities in Fig. 1, i.e., $\{u, m, h, f\}^2$, because it contains a C on both strands of the DNA. We split the transitions of our model into transitions that occur at fixed times and those that occur at random times. This results in a mixture of a discrete time Markov chain (DTMC) and a continuous time Markov chain (CTMC). In the following we will refer to the events or transitions that occur at fixed time points as *discrete part* of the model, while we refer to the other events or transitions at the random time points as *continuous part* of the model.

We assume that cells divide after a fixed time interval (usually every 24 h). Hence, these events correspond to deterministic transitions at fixed times. During cell division one strand is kept as it is (parental strand) and all methylation states and its modifications remain unchanged, while one DNA strand is newly synthesized (daughter strand) and therefore contains only unmethylated cytosine. Consequently, after cell division a CpG that was modified on both sides becomes a CpG that is unmodified on one side. Since the parental strand is chosen at random, the probability for each of the two successor states (corresponding to the state of the CpG in the two daughter cells) is 0.5. The full transition probability matrix $\mathbf{D}$ for cell division is shown in Table 1. Note that the cell division matrix is time-homogeneous.

Maintenance methylation, i.e. methylation events that occur on hemimethylated CpGs to reestablish methylation patterns, is known to be linked to the replication fork [22]. We therefore consider maintenance to occur together with the cell division at the same fixed time points. Hence, cell division and maintenance can be described by a (discrete-time) Markov chain whose transition probability matrix $\mathbf{P}(t)$ is defined in the sequel in Eq. (1). Maintenance may happen at the daughter strand if there is a methylated C on the parental strand, i.e. on hemimethylated CpGs (um or mu), with probability $\mu_m(t)$. Since it is reasonable to assume that hemihydroxylated CpGs (uh or hu) have different properties compared to hemimethylated CpGs in terms of maintaining existing methylation patterns, we describe the maintenance probability for hemihydroxylated CpGs as follows. Let p be the probability that 5hmC is recognized as unmethylated by maintenance enzymes, i.e., the enzyme will not perform maintenance of a hemihydroxylated CpG. Then, the maintenance probability of such a CpG is given by $\bar{p}\mu_m(t)$, where $\bar{p} = 1 - p$. Note that there is no equivalent probability for uf or fu, i.e. we assume that CpGs with 5fC or 5caC at one strand are not maintained [17]. The transition probability matrix for maintenance events is

Table 1. Cell division matrix **D**.

	uu	um	uh	uf	mu	mm	mh	mf	hu	hm	hh	hf	fu	fm	fh	ff
uu	1	0	0	0	0	0	0	0	0	0	0	0	0	0	0	0
um	1/2	1/2	0	0	0	0	0	0	0	0	0	0	0	0	0	0
uh	1/2	0	1/2	0	0	0	0	0	0	0	0	0	0	0	0	0
uf	1/2	0	0	1/2	0	0	0	0	0	0	0	0	0	0	0	0
mu	1/2	0	0	0	1/2	0	0	0	0	0	0	0	0	0	0	0
mm	0	1/2	0	0	1/2	0	0	0	0	0	0	0	0	0	0	0
mh	0	0	1/2	0	1/2	0	0	0	0	0	0	0	0	0	0	0
mf	0	0	0	1/2	1/2	0	0	0	0	0	0	0	0	0	0	0
hu	1/2	0	0	0	0	0	0	0	1/2	0	0	0	0	0	0	0
hm	0	1/2	0	0	0	0	0	0	1/2	0	0	0	0	0	0	0
hh	0	0	1/2	0	0	0	0	0	1/2	0	0	0	0	0	0	0
hf	0	0	0	1/2	0	0	0	0	1/2	0	0	0	0	0	0	0
fu	1/2	0	0	0	0	0	0	0	0	0	0	0	1/2	0	0	0
fm	0	1/2	0	0	0	0	0	0	0	0	0	0	1/2	0	0	0
fh	0	0	1/2	0	0	0	0	0	0	0	0	0	1/2	0	0	0
ff	0	0	0	1/2	0	0	0	0	0	0	0	0	1/2	0	0	0

illustrated in Fig. 2, where we omitted the time dependency of $\mu_m(t)$. Whenever no transition is possible, i.e. there is only a self loop for this state (omitted in Fig. 2) the corresponding diagonal entry in the matrix is 1. Where a transition is possible we set the corresponding (off-diagonal) entry in the matrix to its transition probability and the diagonal entry to 1 minus its transition probability. All other entries in the matrix are 0. One discrete step of the corresponding DTMC corresponds to one cell division, including maintenance methylation. Hence, its transition probability matrix is defined as

$$\mathbf{P}(t) = \mathbf{D} \cdot \mathbf{M}(t). \tag{1}$$

Every other event may occur an arbitrary (unknown) number of times between two cell divisions at random time points and will be described by a continuous-time Markov jump process. These events are *de novo* methylation ($u \rightarrow m$) with rate $\mu_d(t)$, hydroxylation ($m \rightarrow h$) with rate $\eta(t)$, formylation ($h \rightarrow f$) with rate $\phi(t)$ and active demethylation ($f \rightarrow u$) with rate $\delta(t)$ (cf. Fig. 1). Note that all these events may happen on both strands, independent of the state on the complementary strand. All possible reactions are shown in Fig. 3. From these transitions the infinitesimal generator matrix $\mathbf{Q}(t)$ of the jump process can easily be inferred. For the off-diagonal elements, we set the entries to the respective reaction rate if a reaction is possible between two states, indicated by a colored arrow in Fig. 3, and to 0 if no reaction is possible. The diagonal

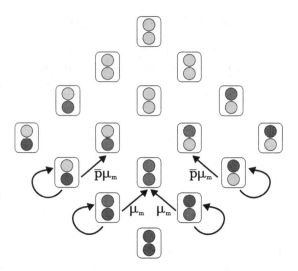

Fig. 2. Maintenance methylation events occur at (fixed) times $t_1, t_2, \ldots, t_n$ and belong therefore to the discrete part of the model. Each state is represented by two colored dots, one for each C on the two strands of the DNA. Unmethylated C is blue, 5mC red, 5hmC yellow and 5fmC cyan. The arrows indicate the possible transitions. Note that we omitted the self loops with probability 1 for states where no transition is possible. The four shown self loops have probability $1 - \mu_m$ or $1 - \bar{p}\mu_m$ respectively. (Color figure online)

elements are then given by the negative sum of the off-diagonal elements of the respective row.

In order to describe the time evolution let us first define the set of time points $T_d = \{t_1, t_2, \ldots, t_n\}$, at which cell division and maintenance occur. We assume that there are in total n of these events. Note that the $t_i \in T_d$ have to be set beforehand, however, one is not restricted to equidistant time intervals. Instead the times can also be sampled from an arbitrary distribution.

At these time points t_i the time evolution of the probability distribution of the states is given by

$$\pi(t_i + \Delta t) = \pi(t_i) \cdot \mathbf{P}(t_i), \tag{2}$$

where Δt is the duration of cell division and maintenance. After cell division and maintenance the other events take place at random time points until the time point for the next cell division is reached. During this interval $[t_i + \Delta t, t_{i+1}]$ the time evolution for $\pi(t)$ is obtained by solving the differential equation

$$\frac{d}{dt}\pi(t) = \pi(t) \cdot \mathbf{Q}(t). \tag{3}$$

Note that since cell division and maintenance methylation occur at a time interval that is much shorter than the time between two cell divisions, we assume here that these events occur instantaneously, i.e. we let $\Delta t \to 0$, which leads to

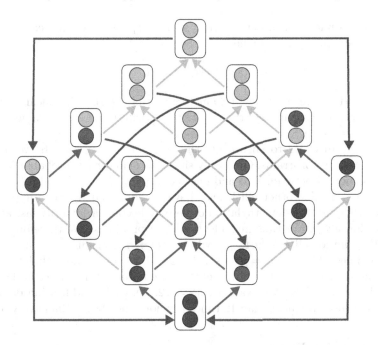

Fig. 3. Possible transitions in the continuous part of the model. Each state is represented by two colored dots, one for each C on the two strands of the DNA. Unmethylated C is blue, 5mC red, 5hmC yellow and 5fmC cyan. The arrows indicate the possible transitions, whereupon the colors indicate the different reactions and their rates, i.e. methylation with rate μ_d (red), oxidation with rate η (yellow), formylation with rate ϕ (cyan) and active demethylation with rate δ (blue). (Color figure online)

a jump of the distribution at t_i. It holds that the left-hand and right-hand limit do not coincide (except for the special case of only uu at time t_i), i.e.

$$\lim_{t \to t_i^-} \pi(t) = \pi(t_i^-) \neq \pi(t_i^-) \cdot \mathbf{P}(t_i^-) = \pi(t_i^+) = \lim_{t \to t_i^+} \pi(t). \qquad (4)$$

To resolve the ambiguity we set the value of π at time t_i to $\pi(t_i) := \lim_{t \to t_i^+} \pi(t)$, which means that we assume that at time t_i the cell division and maintenance methylation have already happened. Intuitively, we can obtain the solution for $\pi(t)$ numerically by alternating between multiplication with $\mathbf{P}$ (Eq. (2)) and integration of Eq. (3).

2.3 Efficiencies

As already discussed in [9] the methylation or modification rates may change over time, i.e. with the assumption of constant rates the behavior of the system can not be captured correctly. We therefore introduce time dependent rates, which we call efficiencies. The simplest way of making the efficiencies time dependent is to impose a linear form. Let $r \in \{\mu_m, \mu_d, \eta, \phi, \delta\}$ be one of the reaction rates. For each reaction we then define

$$r(t) := \alpha_r + \beta_r \cdot t, \tag{5}$$

where α_r and β_r are some new parameters to characterize the linear efficiency function. To further compress the notation we write these new parameters into a vector $\boldsymbol{v}_r = (\alpha_r, \beta_r)$.

Note that in order to ensure identifiability of the parameters we have to introduce certain constrains. Obviously since $\mu_m(t)$ and p are probabilities, they have to be bound between zero and one, i.e. $0 \leq \mu_m(t) \leq 1$ for all $t \in [0, t_{max}]$ and $0 \leq p \leq 1$. The other efficiencies have to be bound by some upper limit ub since otherwise the demethylation cycle may become arbitrarily fast and the optimization algorithm runs into identifiability problems. We therefore require $0 \leq r(t) \leq ub$ for all $t \in [0, t_{max}]$, where $r(t) \neq \mu_m(t)$. Since typical average turnover times $\mathbb{E}[T_{turnover}] = \mu_d^{-1} + \eta^{-1} + \phi^{-1} + \delta^{-1}$ of the demethylation cycle are in the order of 75 to 120 min in certain promotors of human cells [18,25], a viable choice would be, for example, $ub = 12$, i.e. each modification occurs on average not more frequent than 12 times per hour (not faster than every 5 min).

2.4   Conversion Errors

The actual state of a CpG can not be directly observed. We therefore have to estimate the hidden states from sequencing experiments. In total there are 16 hidden und four observable states. Since the different modifications of C might lead to the same observable states it is necessary to perform multiple sequencing experiments in order to uniquely determine the levels of the hidden states. Here, we perform three different kinds of sequencing experiments: All sequencing strategies share a bisulfite treatment, which usually converts C and its oxidized variants 5fC and 5caC, summarized to 5fC*, to uracil. Additionally, in *bisulfite sequencing* (BS) both 5mC and 5hmC remain unconverted while in oxidative bisulfite sequencing (oxBS) only 5mC is retained as C [3]. A combination of both methods can therefore be used to estimate the amount of 5hmC. In *M.SssI assisted bisulfite sequencing* (MAB-Seq) at first all Cs in a CpG context are methylated and afterwards BS is applied. Thus, if all conversions would happen without any errors, only 5fC* would be converted to T [27]. In order to capture the methylation pattern of CpG position at both complementary DNA strands, the distinct chemical treatments were combined with hairpin sequencing [10,11,21]. Regular reactions with their respective probabilities are marked with solid black arrows in Fig. 4, while the possible false reactions are depicted with dashed red arrows. Since every CpG consists of two Cs (one on each strand) with independent conversion errors, we get the conversion error for CpGs by multiplying the individual conversion errors. A complete overview of all possible combinations for each of the three methods is shown in Table 2.

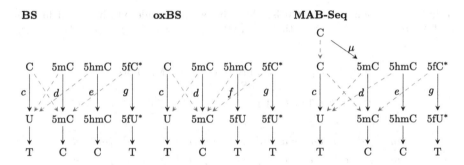

Fig. 4. Cytosine conversions during chemical treatment and sequencing. The correct reactions with their respective probabilities are marked with black arrows, while the false reactions are shown with red dashed arrows. The probability for a false reaction is 1-"rate of correct reaction". (Color figure online)

2.5   Parameter Estimation

Recall that the set of hidden states is $S = \{uu, um, uh, uf, mu, mm, mh, mf, hu, hm, hh, hf, fu, fm, fh, ff\}$. We now define a hidden Markov model (HMM) based on the model presented in Sect. 2.2. As set of observable states we define $S_{obs} = \{TT, TC, CT, CC\}$, i.e. we use the results of the sequencing experiments (cf. Fig. 4) on both strands. The conversion errors define the corresponding emission probabilities. We also define $n_e(j, t)$ as the number of times that state $j \in S_{obs}$ has been observed during independent measurements of sequencing method $e \in E := \{BS, oxBS, MAB\text{-}Seq\}$. The probability distribution over all observable states for experiment e is denoted by $\pi_e(t)$, the probability of a state $j \in S_{obs}$ by $\pi_e(j, t)$ and in a similar fashion we denote $\pi(i, t)$ with $i \in S$ for the hidden states, with probability distribution $\pi(t)$. The observable and hidden states for all times t are connected via

$$\pi_e(t) = \pi(t) \cdot E_e, \tag{6}$$

where E_e is the emission matrix for sequencing method e and is listed in Table 2 for each of the three methods.

Our goal is to estimate the efficiencies for the different methylation events given our hybrid HMM and data from the three different experiments at different time points $t \in T_{obs}$ via a maximum likelihood estimator (MLE). Since an initial distribution over the hidden states, which can not directly be observed, is needed in order to initialize the model, we have to employ the MLE twice: First we estimate the initial distribution $\pi(0)$ over the hidden states by maximizing

$$\pi(0)^* = \mathrm{argmax}_{\pi(0)} \mathcal{L}_1(\pi(0)), \tag{7}$$

under the constrain that $\sum_{i \in S} \pi(i, 0) = 1$. The likelihood $\mathcal{L}_1(\pi(0))$ is defined as

$$\mathcal{L}_1(\pi(0)) = \prod_{e \in E} \prod_{j \in S_{obs}} \pi_e(j, 0)^{n_e(j, 0)}. \tag{8}$$

Table 2. Conversion errors for CpGs, where the rates for single cytosines are defined in Fig. 4. We define $\bar{x} := 1-x$. Note that for MAB-Seq we also define $j := \mu d + (1-\mu)(1-c)$.

	bisulfite seq. (BS)				ox. bisulfite seq. (oxBS)				MAB-Seq			
	TT	TC	CT	CC	TT	TC	CT	CC	TT	TC	CT	CC
uu	c^2	$c\cdot\bar{c}$	$c\cdot\bar{c}$	$\bar{c}^2$	c^2	$c\cdot\bar{c}$	$c\cdot\bar{c}$	$\bar{c}^2$	$\bar{j}^2$	$j\cdot\bar{j}$	$j\cdot\bar{j}$	j^2
um	$c\cdot\bar{d}$	$c\cdot d$	$\bar{c}\cdot\bar{d}$	$\bar{c}\cdot d$	$c\cdot\bar{d}$	$c\cdot d$	$\bar{c}\cdot\bar{d}$	$\bar{c}\cdot d$	$\bar{j}\cdot\bar{d}$	$\bar{j}\cdot d$	$j\cdot\bar{d}$	$j\cdot d$
uh	$c\cdot\bar{e}$	$c\cdot e$	$\bar{c}\cdot\bar{e}$	$\bar{c}\cdot e$	$c\cdot f$	$c\cdot\bar{f}$	$\bar{c}\cdot f$	$\bar{c}\cdot\bar{f}$	$\bar{j}\cdot\bar{e}$	$\bar{j}\cdot e$	$j\cdot\bar{e}$	$j\cdot e$
uf	$c\cdot g$	$c\cdot\bar{g}$	$\bar{c}\cdot g$	$\bar{c}\cdot\bar{g}$	$c\cdot g$	$c\cdot\bar{g}$	$\bar{c}\cdot g$	$\bar{c}\cdot\bar{g}$	$\bar{j}\cdot g$	$\bar{j}\cdot\bar{g}$	$j\cdot g$	$j\cdot\bar{g}$
mu	$c\cdot\bar{d}$	$\bar{c}\cdot\bar{d}$	$c\cdot d$	$\bar{c}\cdot d$	$c\cdot\bar{d}$	$\bar{c}\cdot\bar{d}$	$c\cdot d$	$\bar{c}\cdot d$	$\bar{j}\cdot\bar{d}$	$j\cdot\bar{d}$	$\bar{j}\cdot d$	$j\cdot d$
mm	$\bar{d}^2$	$d\cdot\bar{d}$	$d\cdot\bar{d}$	d^2	$\bar{d}^2$	$d\cdot\bar{d}$	$d\cdot\bar{d}$	d^2	$\bar{d}^2$	$d\cdot\bar{d}$	$d\cdot\bar{d}$	d^2
mh	$\bar{d}\cdot\bar{e}$	$\bar{d}\cdot e$	$d\cdot\bar{e}$	$d\cdot e$	$\bar{d}\cdot f$	$\bar{d}\cdot\bar{f}$	$d\cdot f$	$d\cdot\bar{f}$	$\bar{d}\cdot\bar{e}$	$\bar{d}\cdot e$	$d\cdot\bar{e}$	$d\cdot e$
mf	$\bar{d}\cdot g$	$\bar{d}\cdot\bar{g}$	$d\cdot g$	$d\cdot\bar{g}$	$\bar{d}\cdot g$	$\bar{d}\cdot\bar{g}$	$d\cdot g$	$d\cdot\bar{g}$	$\bar{d}\cdot g$	$\bar{d}\cdot\bar{g}$	$d\cdot g$	$d\cdot\bar{g}$
hu	$c\cdot\bar{e}$	$\bar{c}\cdot\bar{e}$	$c\cdot e$	$\bar{c}\cdot e$	$c\cdot f$	$\bar{c}\cdot f$	$c\cdot\bar{f}$	$\bar{c}\cdot\bar{f}$	$\bar{j}\cdot\bar{e}$	$j\cdot\bar{e}$	$\bar{j}\cdot e$	$j\cdot e$
hm	$\bar{d}\cdot\bar{e}$	$\bar{d}\cdot\bar{e}$	$\bar{d}\cdot e$	$d\cdot e$	$\bar{d}\cdot f$	$d\cdot f$	$\bar{d}\cdot\bar{f}$	$d\cdot\bar{f}$	$\bar{d}\cdot\bar{e}$	$d\cdot\bar{e}$	$\bar{d}\cdot e$	$d\cdot e$
hh	$\bar{e}^2$	$e\cdot\bar{e}$	$e\cdot\bar{e}$	e^2	f^2	$f\cdot\bar{f}$	$f\cdot\bar{f}$	$\bar{f}^2$	$\bar{e}^2$	$e\cdot\bar{e}$	$e\cdot\bar{e}$	e^2
hf	$\bar{e}\cdot g$	$\bar{e}\cdot\bar{g}$	$e\cdot g$	$e\cdot\bar{g}$	$f\cdot g$	$f\cdot\bar{g}$	$\bar{f}\cdot g$	$\bar{f}\cdot\bar{g}$	$\bar{e}\cdot g$	$\bar{e}\cdot\bar{g}$	$e\cdot g$	$e\cdot\bar{g}$
fu	$c\cdot g$	$\bar{c}\cdot g$	$c\cdot\bar{g}$	$\bar{c}\cdot\bar{g}$	$c\cdot g$	$\bar{c}\cdot g$	$c\cdot\bar{g}$	$\bar{c}\cdot\bar{g}$	$\bar{j}\cdot g$	$j\cdot g$	$\bar{j}\cdot\bar{g}$	$j\cdot\bar{g}$
fm	$\bar{d}\cdot g$	$d\cdot g$	$\bar{d}\cdot\bar{g}$	$d\cdot\bar{g}$	$\bar{d}\cdot g$	$d\cdot g$	$\bar{d}\cdot\bar{g}$	$d\cdot\bar{g}$	$\bar{d}\cdot g$	$d\cdot g$	$\bar{d}\cdot\bar{g}$	$d\cdot\bar{g}$
fh	$\bar{e}\cdot g$	$e\cdot g$	$\bar{e}\cdot\bar{g}$	$e\cdot\bar{g}$	$f\cdot g$	$\bar{f}\cdot g$	$f\cdot\bar{g}$	$\bar{f}\cdot\bar{g}$	$\bar{e}\cdot g$	$e\cdot g$	$\bar{e}\cdot\bar{g}$	$e\cdot\bar{g}$
ff	g^2	$g\cdot\bar{g}$	$g\cdot\bar{g}$	$\bar{g}^2$	g^2	$g\cdot\bar{g}$	$g\cdot\bar{g}$	$\bar{g}^2$	g^2	$g\cdot\bar{g}$	$g\cdot\bar{g}$	$\bar{g}^2$

Note that Eq. (8) is independent of the parameters. Given an initial distribution over the hidden states we can now run our model and apply the MLE

$$v^* = \operatorname{argmax}_v \mathcal{L}_2(v), \tag{9}$$

a second time in order to estimate the efficiencies, where

$$\mathcal{L}_2(v) = \prod_{e\in E}\prod_{t\in T_{obs}\setminus\{0\}}\prod_{j\in\mathcal{S}_{obs}} \pi_e(j,t)^{n_e(j,t)}. \tag{10}$$

The vector $v = (v_{\mu_m}, v_{\mu_d}, v_\eta, v_\phi, v_\delta, p)$ contains all unknown parameters for all efficiencies and the probability p of considering a hydroxylated cytosine as unmethylated. Note that applying the MLE twice and independently leads only to an approximation of the true most likely explanation, since the estimated initial distribution may not lead to the same result in the parameter estimation, as if it would all be done in one estimation. However, we choose this approach to reduce the computational complexity of the optimization. Note that we used a numerical multistart optimization approach for both MLEs in order to ensure that the global optimum is indeed found.

In order to estimate the standard deviations of the estimated parameters we use the observed Fisher information matrix [5]. The Fisher information is defined

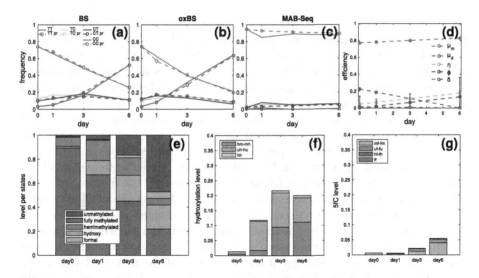

Fig. 5. Results for Afp. (a)–(c) Predicted frequencies (dashed lines) and frequencies obtained from sequencing experiments (solid lines) of the observable states for all three methods. (d) Estimated efficiencies with standard deviations. (e) Probabilities of the hidden states. (f) Detailed distribution of hydroxylated CpGs. (g) Detailed distribution of formalized CpGs.

as $\mathcal{J}(v^*) = -\mathcal{H}(v^*)$, where v^* is the maximum likelihood estimate and $\mathcal{H}(v) = \nabla\nabla^T \log \mathcal{L}_2(v)$ the Hessian matrix of the log-likelihood. The expected Fisher information is then given by $\mathcal{I}(v^*) = \mathbb{E}[\mathcal{J}(v^*)]$ and its inverse forms a lower bound for the covariance matrix. Thus, we can approximate the standard deviation of all estimated parameters by $\sigma(v^*) = \sqrt{\mathrm{Var}(v^*)} \approx \sqrt{\mathrm{diag}(-\mathcal{H}^{-1}(v^*))}$.

The implementation of the hybrid HMM and its analysis as explained above has been integrated into the latest beta version of the H(O)TA tool [20]. H(O)TA provides results for individual CpGs and also an aggregated profile across all analyzed CpGs.

3   Results

In the following, we will discuss the results after applying our model to data derived from a short region at the single copy gene Afp (alpha fetoprotein), which contains 5 CpGs. More precisely, we followed the DNA methylation changes during the adjustment of mouse embryonic stem cells (mESCs) towards 2i medium after previous long time cultivation under Serum/LIF conditions [8,9]. The Serum/ LIF-to-2i shift is a common model system which induces genome wide demethylation in mESCs including the Afp locus.

The bases of our modelling is given by three data sets derived from three sequencing experiments, hairpin (HP) bisulfite sequencing (BS), HP oxidative BS (HPoxBS) and M.SssI assisted HPBS (MAB-Seq). While the combination of

BS and oxBS permits the simultaneous detection of 5mC and 5hmC, MAB-Seq provides the combined level and distribution of 5fC and 5caC [3,10,11,21,27]. Thus, the additional information about oxidized cytosine forms allows us to investigate the role of 5hmC and 5fC/5caC in the given demethylation process.

Here, the individual and aggregated H(O)TA results show a very similar behavior. Hence, we only present the aggregated results in Fig. 5.

Figure 5 (a)–(c) shows a comparison of the actual measurements (solid lines) and the results from the model (dashed lines) of the time evolution for the four observable states for all three sequencing models. The predictions and measurements are in good agreement. In BS and oxBS over time the frequency of TT increases, while the frequency of CC decreases. Moreover, the frequencies of TC and CT show a temporal increase. For MAB-Seq the frequencies of all observable states remain quite constant, except for some small changes in early times. This behavior can be explained by the changes over time in the efficiencies, which are shown in Fig. 5(d). The maintenance probability μ_m remains constant at about 0.8, as well as the probability p of considering 5hmC as unmethylated, which is constant by definition. The estimation of p gives $p = 1$, i.e. 5hmC is always considered to be unmethylated and will not be recognized by Dnmt1 after replication which means that 5hmC leads to an impairment of Dnmt1 activity and a passive loss of DNA methylation with each replication. The *de novo* efficiency μ_d decreases over time, while the hydroxylation and formylation efficiency η and ϕ increase. The demethylation efficiency δ is always 0. Note, however, that the standard deviation for δ becomes very large for later time points due to insufficient data. On the contrary the standard deviations for all other efficiencies remain very small.

The efficiencies explain the behavior of the frequencies of the observable states and is even more evident for the hidden states shown in Fig. 5(e): Over time the probability of being fully or hemimethylated decreases, while the probability of being unmethylated, hydroxylated or formylated increases. A more detailed look into hydroxylated and formylated states is shown in Fig. 5(f) and (g). Note that the combination of hydroxylation and formylation is only shown in one of the two subplots, namely in (g). The observed increase in 5hmC and 5fC/5caC is in accordance with the high oxidation efficiency of Test in form of hydroxylation and formylation. Previously, we showed using a purely discrete HMM, that the presence of 5hmC leads to a block of Dnmt1 activity after replication, which we also observe in the present hybrid model [9]. However, we now also observe an increase in higher oxidized cytosine variants, namely 5fC and 5caC which equally prevents methylation by Dnmt1. Thus, we reason that the impact of Tet mediated oxidation of 5mC on DNA demethylation in the investigated system plays a much more important role than previously suggested [26].

Considering the rather rapid periodic events of *de novo* methylation and active demethylation, the chosen measurement time points are not ideal. There is only information available at the end of each cell division cycle (one division within 24 h), i.e., no information is given for the times between two cell divisions. With measurements at time points between two cell divisions we would be able to distinguish if a CpG is in a certain state because it initially was, or because it ran through the full cycle, possibly multiple times. We emphasize that with better data, i.e. with sufficiently many measurements between cell divisions, it will be possible to estimate all reaction efficiencies with better confidence.

4   Conclusion

We proposed a hybrid hidden Markov model which is able to successfully describe both, events such as cell division and maintenance methylation that occur at fixed times, and events that occur at random times, such as *de novo* methylation, oxidizations and active demethylation, according to a continuous-time Markov jump process. To the best of our knowledge, this is the first model that describes the dynamics of active demethylation, i.e., the active removal of the methyl group through several enzymatic steps. We applied our model to data from mouse embryonic stem cells, which undergo a gradually loss of DNA methylation over time. We were able to accurately predict the frequency of the observable states and the levels of the hidden states in all cases. We were also able to predict the enzymatic reaction efficiencies based on a linear assumption for their time behavior.

As future work we plan to apply our model to more informative data such that all efficiencies of the active demethylation cycle can be estimated with better confidence. Moreover, we plan to allow different functional forms for the efficiencies as we do not always expect a constant in- or decrease in enzyme efficiencies over time. Therefore, a linear form is not flexible enough and does not allow to capture more complex behaviors. A suitable choice could be splines of different degrees and with a different number of knots. However, in this case it is also necessary to perform model selection in order to prevent overfitting. Another possible extension could be to investigate potential neighborhood dependencies of the modified cytosines [23].

References

1. Arand, J., et al.: In vivo control of CpG and non-CpG DNA methylation by DNA methyltransferases. PLoS Genet. **8**(6), e1002750 (2012)
2. Bestor, T.H., Ingram, V.M.: Two DNA methyltransferases from murine erythroleukemia cells: purification, sequence specificity, and mode of interaction with DNA. Proc. Nat. Acad. Sci. **80**(18), 5559–5563 (1983)
3. Booth, M.J., et al.: Quantitative sequencing of 5-methylcytosine and 5-hydroxymethylcytosine at single-base resolution. Science **336**(6083), 934–937 (2012)
4. Cardoso, M.C., Leonhardt, H.: DNA methyltransferase is actively retained in the cytoplasm during early development. J. Cell Biol. **147**(1), 25–32 (1999)
5. Efron, B., Hinkley, D.V.: Assessing the accuracy of the maximum likelihood estimator: observed versus expected fisher information. Biometrika **65**(3), 457–483 (1978)

6. Ehrlich, M., et al.: Amount and distribution of 5-methylcytosine in human DNA from different types of tissues or cells. Nucleic Acids Res. **10**(8), 2709–2721 (1982)
7. Feng, S., et al.: Conservation and divergence of methylation patterning in plants and animals. Proc. Nat. Acad. Sci. **107**(19), 8689–8694 (2010)
8. Ficz, G., et al.: FGF signaling inhibition in ESCs drives rapid genome-wide demethylation to the epigenetic ground state of pluripotency. Cell Stem Cell **13**(3), 351–359 (2013)
9. Giehr, P., Kyriakopoulos, C., Ficz, G., Wolf, V., Walter, J.: The influence of hydroxylation on maintaining CpG methylation patterns: a hidden Markov model approach. PLoS Comput. Biol. **12**(5), 1–16 (2016)
10. Giehr, P., Kyriakopoulos, C., Lepikhov, K., Wallner, S., Wolf, V., Walter, J.: Two are better than one: HPoxBS-hairpin oxidative bisulfite sequencing. Nucleic Acids Res. **46**(15), e88 (2018)
11. Giehr, P., Walter, J.: Hairpin bisulfite sequencing: synchronous methylation analysis on complementary DNA strands of individual chromosomes. In: Tost, J. (ed.) DNA Methylation Protocols. MMB, vol. 1708, pp. 573–586. Springer, New York (2018). https://doi.org/10.1007/978-1-4939-7481-8_29
12. Globisch, D., et al.: Tissue distribution of 5-hydroxymethylcytosine and search for active demethylation intermediates. PLoS ONE **5**(12), e15367 (2010)
13. Hashimoto, H., et al.: Recognition and potential mechanisms for replication and erasure of cytosine hydroxymethylation. Nucleic Acids Res. **40**(11), 4841–4849 (2012)
14. He, Y.F., et al.: Tet-mediated formation of 5-carboxylcytosine and its excision by TDG in mammalian DNA. Science **333**(6047), 1303–1307 (2011)
15. Hermann, A., Goyal, R., Jeltsch, A.: The Dnmt1 DNA-(cytosine-C5)-methyltransferase methylates DNA processively with high preference for hemimethylated target sites. J. Biol. Chem. **279**(46), 48350–48359 (2004)
16. Ito, S., et al.: Tet proteins can convert 5-methylcytosine to 5-formylcytosine and 5-carboxylcytosine. Science **333**(6047), 1300–1303 (2011)
17. Ji, D., Lin, K., Song, J., Wang, Y.: Effects of Tet-induced oxidation products of 5-methylcytosine on Dnmt1-and DNMT3a-mediated cytosine methylation. Mol. BioSyst. **10**(7), 1749–1752 (2014)
18. Kangaspeska, S., et al.: Transient cyclical methylation of promoter DNA. Nature **452**(7183), 112 (2008)
19. Kriaucionis, S., Heintz, N.: The nuclear DNA base 5-hydroxymethylcytosine is present in Purkinje neurons and the brain. Science **324**(5929), 929–930 (2009)
20. Kyriakopoulos, C., Giehr, P., Wolf, V.: H(O)TA: estimation of DNA methylation and hydroxylation levels and efficiencies from time course data. Bioinformatics **33**(11), 1733–1734 (2017)
21. Laird, C.D., et al.: Hairpin-bisulfite PCR: assessing epigenetic methylation patterns on complementary strands of individual DNA molecules. Proc. Nat. Acad. Sci. **101**(1), 204–209 (2004)
22. Leonhardt, H., Page, A.W., Weier, H.U., Bestor, T.H.: A targeting sequence directs DNA methyltransferase to sites of DNA replication in mammalian nuclei. Cell **71**(5), 865–873 (1992)
23. Lück, A., Giehr, P., Walter, J., Wolf, V.: A stochastic model for the formation of spatial methylation patterns. In: Feret, J., Koeppl, H. (eds.) CMSB 2017. LNCS, vol. 10545, pp. 160–178. Springer, Cham (2017). https://doi.org/10.1007/978-3-319-67471-1_10

24. Maiti, A., Drohat, A.C.: Thymine DNA glycosylase can rapidly excise 5-formylcytosine and 5-carboxylcytosine potential implications for active demethylation of CpG sites. J. Biol. Chem. **286**(41), 35334–35338 (2011)

25. Métivier, R., et al.: Cyclical DNA methylation of a transcriptionally active promoter. Nature **452**(7183), 45 (2008)

26. von Meyenn, F., et al.: Impairment of DNA methylation maintenance is the main cause of global demethylation in naive embryonic stem cells. Mol. Cell **62**(6), 848–861 (2016)

27. Neri, F., Incarnato, D., Krepelova, A., Parlato, C., Oliviero, S.: Methylation-assisted bisulfite sequencing to simultaneously map 5fC and 5caC on a genome-wide scale for DNA demethylation analysis. Nat. Protoc. **11**(7), 1191 (2016)

28. Okano, M., Bell, D.W., Haber, D.A., Li, E.: DNA methyltransferases Dnmt3a and Dnmt3b are essential for de novo methylation and mammalian development. Cell **99**(3), 247–257 (1999)

29. Okano, M., Xie, S., Li, E.: Cloning and characterization of a family of novel mammalian DNA (cytosine-5) methyltransferases. Nat. Genet. **19**(3), 219 (1998)

30. Tahiliani, M., et al.: Conversion of 5-methylcytosine to 5-hydroxymethylcytosine in mammalian DNA by MLL partner TET1. Science **324**(5929), 930–935 (2009)

Using a Hybrid Approach to Model Central Carbon Metabolism Across the Cell Cycle

Cecile Moulin[1,2]([✉]), Laurent Tournier[2], and Sabine Peres[1,2]

[1] LRI, Université Paris-Sud, CNRS, Université Paris-Saclay, 91405 Orsay, France
moulin@lri.fr
[2] MaIAGE, INRA, Université Paris-Saclay, 78350 Jouy-en-Josas, France

Abstract. Metabolism and cell cycle are two central processes in the life of a eukaryote cell. If they have been extensively studied in their own right, their interconnection remains relatively poorly understood. In this paper, we propose to use a differential model of the central carbon metabolism. After verifying the model accurately reproduces known metabolic variations during the cell cycle's phases, we extend it into a hybrid system reproducing an imposed succession of the phases. This first hybrid approach qualitatively recovers observations made in the literature, providing an interesting first step towards a better understanding of the crosstalks between cell cycle and metabolism.

Keywords: Metabolism · Cell cycle · Ordinary differential equations · Hybrid systems

1 Introduction

In cell biology, two mechanisms are absolutely central to understand the growth of a cell population: the metabolism and the cell division cycle. The former deals with the production of energy and the production of all molecular components needed for a cell to live and grow, while the latter ensures, when a cell has sufficiently grown, that it will divide into two daughter cells. Both mechanisms have been extensively studied over the years, however the exploration of inter-dependency between them is relatively recent [2,10,15]. The coupling of the two systems has attracted a lot of attention in recent years, as disruptions in their interconnection have been linked to severe pathologies such as cancer [3].

Similarly, from a modeling point of view the dynamical analysis of metabolism (*e.g.* [12]) and of cell cycle (*e.g.* [7]) are also relatively separate, and the coupling of metabolic and cell cycle models remains a challenging task in systems biology. In this paper, we propose a first approach towards this coupling by proposing a differential model of central carbon metabolism (CCM), inspired from [4]. The CCM is an important metabolic part regarding both synthesis of precursors (amino acids, nucleotides, fatty acids *etc*), energy production (ATP) or redox ratios (NAD/NADH, NADP/NADPH), making this model

© Springer Nature Switzerland AG 2019
M. Češka and N. Paoletti (Eds.): HSB 2019, LNBI 11705, pp. 132–146, 2019.
https://doi.org/10.1007/978-3-030-28042-0_9

a good candidate to analyze the effect of cell cycle on metabolic activity. We then construct different versions of the CCM model, reproducing major known effects of the cycle's phases. Finally we use a hybrid approach to simulate the dynamical succession of the phases.

This article is motivated by previous works of da Veiga Moreira *et al.* [3], where experimental results were obtained, in normal and cancer cells. In particular, they measured ATP concentrations and redox ratios at different phases of the cell cycle. These measures highlight variations across the cycle, with notable differences between normal and cancer cells. By qualitatively reproducing key experimental observations in normal cells in [3], our hybrid dynamical model constitutes an important first step towards a better understanding of metabolism along the cell cycle.

This article is organized as follows. We first present the dynamical model of the CCM (Sect. 2). Then, we analyze its sensitivity with respect to known regulations of the cell cycle, and propose different versions of the model, one for each phase (Sect. 3). Finally, we combine them into a hybrid model and simulate it on a full cycle (Sect. 4).

2   A Dynamical Model of Central Carbon Metabolism

We start with a brief presentation of the metabolic model used in this paper. This model consists of 24 ordinary differential equations summarizing the main pathways, together with the main regulations of the central carbon metabolism (CCM) in a growing population of eukaryotic cells. It is a slight adaptation of the model in [4], which is itself based on the work of Robitaille [12].

2.1   Description of the Metabolic Model

The state vector $x(t) \in \mathbb{R}_+^{24}$ is decomposed into three groups of variables:

$$x(t) = (x_I(t), x_{II}(t), x_{biomass}(t)),$$

where $x_I \in \mathbb{R}_+^{16}$ contains the concentration (expressed in mmol L^{-1}) of 16 metabolites in the CCM, including palmitate to represent lipid production; $x_{II} \in \mathbb{R}_+^7$ contains the concentration (expressed in mmol L^{-1}) of central cofactors: ATP-ADP-AMP (energy management), NAD-NADH and NADP-NADPH (redox management); finally $x_{biomass} \in \mathbb{R}_+$ represents biomass production, it is expressed in L. The differential system can be summarized by:

$$\begin{cases} \dot{x}_I & = \mathcal{S}_I \nu(x(t)) - \mu(x(t)) x_I(t), \\ \dot{x}_{II} & = \mathcal{S}_{II} \nu(x(t)), \\ \dot{x}_{biomass} & = \mu(x(t)) x_{biomass}(t). \end{cases} \tag{1}$$

Evolution of Metabolites. The first two equations in (1) govern the temporal evolution of the different metabolites and cofactors. The term $-\mu x_I$ represents a dilution term, modeling the fact that the cell population is growing. Note that the second equation does not have a dilution term: it is a modeling assumption to take into account the fact that the biosynthetic pathways of cofactors (such as the couple NAD/NADH for instance) are not included in the model. More precisely: every time a molecule of NADH is produced in the model, a molecule of NAD is consumed so that the sum $y(t) = NAD(t) + NADH(t)$ keeps constant. Thus, since the production *de novo* of these "exchange" metabolites is not included in the model (for the sake of simplicity), their dilution is ignored preventing unwanted exhaustion of their pools. This is the only reason why the metabolites are separated into the two groups of variables x_I and x_{II}. The remainder of the right-hand terms can be regrouped:

$$\begin{pmatrix} \mathcal{S}_I \\ \mathcal{S}_{II} \end{pmatrix} \nu(x(t)) = \mathcal{S}\nu(x(t)),$$

where $\mathcal{S} \in \mathbb{Q}^{23 \times 29}$ is the stoichiometric matrix and $\nu(x) \in \mathbb{R}^{29}$ is the vector of velocities of the 29 reactions involved in the model. These reactions recover the main pathways of the CCM: glycolysis, pentose-phosphate pathway (PPP), tricarboxylic acid cycle (TCA) as well as lactate and lipid production (see Fig. 1 for a graphical representation). Note that some reactions have been aggregated, according to the original model in [12].

The construction of the vector $\nu(x(t))$ of reaction rates is mainly based on the classical model of Michaelis-Menten for enzymatic catalysis. According to this model, the velocity of reaction $S \xrightarrow{E} P$ is given by

$$\nu = k_{cat}e(t)\frac{s(t)}{K_m + s(t)}, \tag{2}$$

where $e(t)$ and $s(t)$ designate the concentrations in enzyme and substrate, and k_{cat} and K_M are constants. For a reversible reaction r, we split it into a forward and a backward reaction:

$$\nu_r = \nu_{r\_f} - \nu_{r\_b},$$

the velocity of each reaction being given by (2), with different k_{cat} and K_m. In our model, enzyme concentrations are supposed fixed[1], simplifying (2) into

$$\nu = \nu_{max}\frac{s}{K_m + s}, \quad \text{where} \quad \nu_{max} = k_{cat}e. \tag{3}$$

Finally, some known regulatory effects are included in the model, mainly under the form of multiplicative factors involving michaelis-like terms. To better see how the differential equations are constructed, let us consider an example.

---

[1] Note that this fact amounts to supposing that the "genetic part" of the cell ensures the maintenance of enzymatic pools.

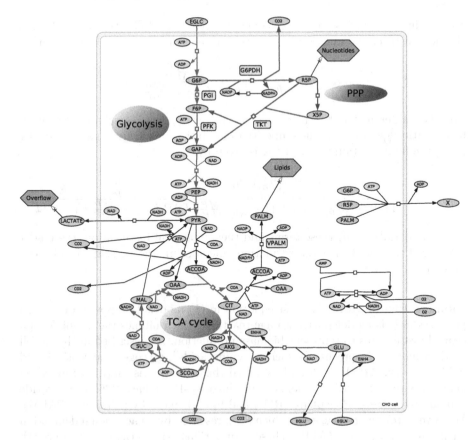

Fig. 1. Graphical representation of the CCM model designed with Celldesigner [8]. Glycolysis is represented in red, pentose phosphate pathway in blue and TCA cycle in green. Reversible reactions are represented with a double arrow, as for instance PGI and irreversible reactions by a simple arrow. Note that some reactions have been aggregated (following [12]; for instance PFK denotes the aggregation of phosphofructokinase, aldolase and triose phosphate isomerase). (Color figure online)

Take for instance the fructose 6-phosphate $F6P$, which is the second metabolite of the glycolysis. By looking at the different reactions involving F6P (see Fig. 1), its differential equation reads:

$$\frac{dF6P}{dt} = \nu_{pgi\_f} - \nu_{pgi\_b} - \nu_{pfk} + 2\nu_{tkt} - \mu F6P.$$

Now, let us explain how the reaction rates ν are built. Consider for example the term ν_{pfk}, which is the rate of phosphofructokinase. In the model, this reaction is actually aggregated with two neighbor reactions (catalyzed by fructose biphosphate aldolase and triose phosphate isomerase), leading to the overall reaction:

$$F6P + ATP \longrightarrow 2GAP + ADP.$$

To take into account the effect of cofactors ADP and ATP, we follow [12] and [4] and construct the function ν_{pfk} as the product of Michaelis-Menten terms:

$$\nu_{pfk} = \nu_{max} \frac{F6P}{K_{m1} + F6P} \frac{\frac{ATP}{ADP}}{K_{m2} + \frac{ATP}{ADP}}.$$

Now, two different regulations of this enzyme are known, namely a stimulation by the ratio $\frac{AMP}{ATP}$ and a (non-competitive) inhibition by citrate. Again, following [4,12] the final reaction rate of PFK is given by:

$$\nu_{pfk} = \nu_{max} \frac{\left(1 + \frac{\beta}{\alpha K} \frac{AMP}{ATP}\right) F6P}{\left(1 + \frac{1}{K} \frac{AMP}{ATP}\right) K_{m1} + \left(1 + \frac{1}{\alpha K} \frac{AMP}{ATP}\right) F6P} \frac{\frac{ATP}{ADP}}{K_{m2} + \frac{ATP}{ADP}} \frac{K_i}{K_i + CIT}.$$

For more details about these terms and their biological justification, the reader is referred to [12,13]. All the regulations included in the present model come from [4].

Biomass Production. The last equation in (1) models the growth of the cell population. It is modeled as a storage reaction, which mimics the use of certain central metabolites to represent biomass production. As described in [14], a cell needs a number of precursors such as amino acids[2], glycogen, nucleotides and lipids to grow. Articles [12] and [4] make the following assumptions: glycogen demand is represented by a demand in glucose-6-Phosphate (G6P), nucleotide demand by ribose-5-Phosphate (R5P) and lipid demand by palmitate (PALM). Moreover, growth needs energy, which is represented by an additional demand in adenosine triphosphate (ATP). These assumptions lead to the following growth rate:

$$\mu(x(t)) = \nu_{max} \frac{ATP}{K_{ATP} + ATP} \frac{G6P}{K_{G6P} + G6P} \frac{R5P}{K_{R5P} + R5P} \frac{PALM}{K_{PALM} + PALM}. \tag{4}$$

2.2   Simulation of the Model

The model consists in 24 highly nonlinear differential equations. It involves around a hundred parameters, which mainly come from [12] and [4]. Given its complexity, an analytical analysis of its asymptotic behavior is presently not available. Nevertheless, we performed multiple numerical simulations to test its behavior in different conditions. With respect to [12], we further simplified by considering glucose as the unique source of carbon (in the model, external glucose directly enters as G6P through the hexokinase reaction). In general, the model seems to converge to a stationary regime illustrated in Fig. 2.

---

[2] For the sake of simplicity, the demands in amino acids are ignored in this paper.

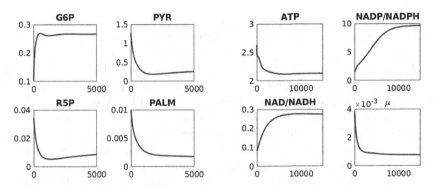

Fig. 2. Example of temporal trajectory of the metabolic model. On the left, the metabolites visually reach a stationary regime around 5 000 h. On the right, ATP, NAD/NADH and μ reach a stationary regime around 10 000 h while the ratio NADP/NADPH needs more time to reaches it. All numerical simulations of this paper are made in Matlab (The MathWorks, Inc.) with solver ode15s.

Such a stationary regime is not a proper equilibrium point, as the biomass grows exponentially. However, it seems to represent an equilibrium regime modeling a population growing in exponential phase, where all internal metabolites numerically tend to a steady state value. As expected in such conditions, glycolysis is fully running (glucose 6-phosphate and pyruvate are respectively the entry and exit points of glycolysis), while pentose phosphate pathway and lipid production are relatively low (represented respectively by ribose 5-phosphate and palmitate).

An advantage of this model is that it explicitly includes the concentration of "exchange" metabolites (group II) as variables. It is therefore possible to have a direct visualization of key functions of the cell such as energy production (through ATP) or its redox state whether in catabolism (through the ratio NAD/NADH) or in anabolism (through the ratio NADP/NADPH).

3   Reproducing Metabolic Variations Along the Cell Cycle

The system presented in the previous section is intended to model the temporal evolution of a (exponentially) growing population of cells. For a population to grow, cells need to divide, undergoing the complex process known as *cell cycle* (or cell-division cycle). This process can be defined as the series of key cellular events, including the duplication of the genetic material, to ultimately produce two genetically identical daughter cells. Schematically, it can be decomposed into four successive phases: G1-S-G2-M (sometimes a fifth phase G0 is added representing a quiescent state). The regulatory network behind this process has been extensively studied in different biological contexts, and models using different formalisms are available [7,11]. However, understanding the interconnection between the cell cycle and metabolism remains challenging.

To understand the links between the two, a key is to understand the main role of each phase in the division process. The effects of the cycle on metabolic activity have been actively explored recently, leading to both experimental and modeling works (see *e.g.* [1,3,5,6] and references therein). Phase G1 is typically viewed as a growth phase, where the cell produces a lot of proteins. It is generally associated with high energy production, notably as ATP. Also associated with high energy, phase S is the phase of DNA duplication; it is linked to an increase of the pentose phosphate pathway, main producer of nucleotide precursors. During G2 the cell continues its growth, notably by producing membrane precursors (lipids in particular). Finally, phase M is the mitosis phase itself, where the cell actually divide into two daughter cells. Little is known about specific metabolic activity in this phase. Table 1 summarizes the main metabolic variations through the cycle's phases.

Table 1. Qualitative description of the main role of mammalian cell cycle's phases and specific metabolite demands along the phases (mainly interpreted from [3,5]).

	Main role	Specific metabolic demands
G1	Growth, production of proteins (and mRNA)	Energy (ATP) amino acids, nucleotides
S	DNA duplication	Energy and nucleotides
G2	Growth, membrane production	lipids
M	Division	not known

The main idea of this article is to use the metabolic model described in Sect. 2 to reproduce, at least on a qualitative level, the general behavior of metabolism during the cycle's phases, as described in Table 1. As a first step, in the following we investigate whether the model is able to reproduce major regulations of the CCM by the cell cycle. For that, we mainly use the review article [6] by Diaz-Moralli *et al.*, which outlines some of the major known regulations. After verifying that the model is able to reproduce all major effects described in [6], we further use this result to build three different versions of the model, each representing the CCM during a phase of the cell cycle.

3.1   Uncovering Regulatory Effects of the Cycle Phases on the CCM

For each phase G1, S and G2 we proceed in the same way. The first step is to deduce from [6] the main regulatory effects of the phase on CCM enzymes. Table 2 below reproduces all tested effects. Each regulatory effect may be positive or negative, and can act on enzyme activity (*e.g.* increased activity of G6PDH in S) or directly on enzyme level (*e.g.* accumulation of PFK in G1). In both cases, we proceed in the same way and make vary the parameter ν_{max} (maximal velocity, see Eq. (3) above): an increase of ν_{max} either corresponds to an increased activity (*i.e.* a higher k_{cat}) or to an increase of enzyme concentration (a higher e). The second step is to select a "response" metabolite x in the model to test the expected effect. Sometimes the choice is straightforward; for instance in G1 the increase of lipogenic activity is supposed to have an effect on lactate concentration, so we test $x = LAC$. Sometimes the choice is more indirect; for instance in S the decrease of PFK concentration is linked with an increase of the pentose phosphate pathway, in that case we choose $x = R5P$ (ribose 5-phosphate, in the middle of the PPP, see Fig. 1). Once the enzyme and the response metabolite x are chosen, we simulate the model and plot the steady state value x^* for different value of the enzyme ν_{max}, thus allowing to test whether the model reproduce the desired (positive or negative) effect.

Table 2. List of the tested effects of cell cycle's phases on the CCM, as interpreted from [6]. Note that **VPALM** corresponds to the reaction of lipid production (palmitate in our case). In G1, PFK has a high concentration while G6PDH has a low activity.

Enzyme	G1	S	G2
PFK	High	Low	Low
G6PDH	Low	High	High
TKT	Low	Low	High
VPALM	Low	Low	High

In total, we tested six different effects indicated in [6] and in each case the model recovered the expected behavior (increase or decrease of the response metabolite with respect to varying ν_{max}). These experiments are described thereafter. Since these results were encouraging, we decided to go further and to instantiate the model into three versions, each one representing a phase of the cycle. These versions were simply deduced by arbitrarily choosing high or low values for the parameters ν_{max} for each of the four enzymes PFK, G6PDH, TKT and VPALM.

Phase G1. According to [6], two main enzymes are affected during G1. First, there is an accumulation of PFK during the whole phase, leading to an increase in glycolytic activity. To observe such an activity, we tracked the concentration of pyruvate, which is the end-product of glycolysis. Figure 3 clearly shows an increase of PYR* with respect to ν_{maxPFK}. The second regulation in G1 indicated by [6] is a decrease of lipogenic enzyme concentration leading to an increase of lactate concentration. In our model, we used the enzyme VPALM (production of palmitate) to test a decrease of lipid production, and we observed the steady state concentration of lactate. Again, as expected the model reproduce an increase of lactate when $\nu_{maxVPALM}$ decreases (see Fig. 3). Therefore, to obtain a model for G1 we set the two parameters to the following values:

$$\begin{cases} \nu_{\text{maxPFK}}^{\text{G1}} := 1 \cdot 10^{-2} \text{ mmol } L^{-1} h^{-1}, \\ \nu_{\text{maxVPALM}}^{\text{G1}} := 3 \cdot 10^{-6} \text{ mmol } L^{-1} h^{-1}. \end{cases}$$

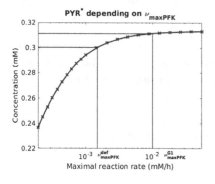 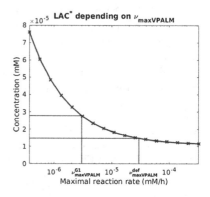

Fig. 3. Left: steady state value of pyruvate concentration for varying values of ν_{maxPFK}. Right: steady state value of lactate concentration for varying values of ν_{maxVPALM}. Values $\nu_{\text{max}}^{\text{def}}$ and $\nu_{\text{max}}^{\text{G1}}$ correspond respectively to the value of the parameter ν_{max} in the initial model and in the G1 model.

Phase S. During this phase the concentration of PFK starts to decrease [6], leading to a subsequent increase of the PPP. As indicated earlier, we decided to observe this effect on the concentration of R5P, which in the model is the central metabolite of the PPP. The second change in this phase is an increase of G6PDH activity, also contributing to an increase of R5P. Figure 4 shows that the model reproduces both effects. Furthermore, since in phase S lipogenic enzymes are still supposed to be low, we instanciate a S model by setting:

$$\begin{cases} \nu_{\text{maxVPALM}}^{\text{S}} := \nu_{\text{maxVPALM}}^{\text{G1}}, \\ \nu_{\text{maxPFK}}^{\text{S}} := 1.5 \cdot 10^{-3} \text{ mmol } L^{-1} h^{-1}, \\ \nu_{\text{maxG6PDH}}^{\text{S}} := 5 \cdot 10^{-4} \text{ mmol } L^{-1} h^{-1}. \end{cases}$$

Phase G2. In G2 the concentration of lipogenic enzymes increases, leading to an increase of lipid production. Lipids in the model are represented by palmitate. Moreover, increase in the activity of TKT is supposed to further activate the end of the PPP [6]. We used the metabolite GAP in the model, as it is the exit point of the PPP (where the PPP flux comes back to glycolysis). Both effects are qualitatively reproduced in the model (see Fig. 5). Further taking into account

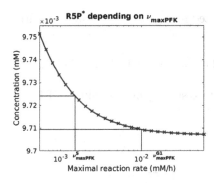
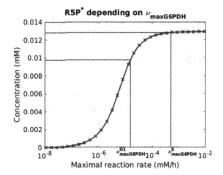

Fig. 4. Left: steady state R5P concentration with respect to ν_{maxPFK} with $\nu_{\mathrm{maxVPALM}}^{\mathrm{S}}$ set. Right: steady state R5P concentration with respect to ν_{maxG6PDH}.

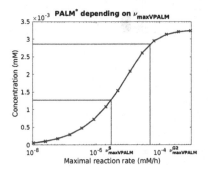
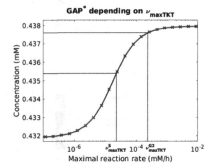

Fig. 5. Left: steady state PALM concentration with respect to ν_{maxVPALM} with $\nu_{\mathrm{maxG6PDH}}^{\mathrm{G2}}$ and $\nu_{\mathrm{maxPFK}}^{\mathrm{G2}}$ set. Right: steady state R5P concentration with respect to ν_{maxG6PDH}.

the fact that PFK concentration is supposed to be low and the activity of G6PDH is supposed to be high, we instantiate a model G2 by setting:

$$\begin{cases} \nu_{\mathrm{maxPFK}}^{\mathrm{G2}} & := \nu_{\mathrm{maxPFK}}^{\mathrm{S}}, \\ \nu_{\mathrm{maxG6PDH}}^{\mathrm{G2}} & := \nu_{\mathrm{maxG6PDH}}^{\mathrm{S}}, \\ \nu_{\mathrm{maxVPALM}}^{\mathrm{G2}} & := 5 \cdot 10^{-5} \ \mathrm{mmol} \ \mathrm{L}^{-1} \ \mathrm{h}^{-1}, \\ \nu_{\mathrm{maxTKT}}^{\mathrm{G2}} & := 2.3 \cdot 10^{-4} \ \mathrm{mmol} \ \mathrm{L}^{-1} \ \mathrm{h}^{-1}. \end{cases}$$

3.2   Validation of the Three Models

Thanks to biological information [6], we were thus able to propose three versions of the original model, each supposed to reproduce the main metabolic behavior induced during one of the cycle's phases G1, S and G2. To further validate those models, we simulated them separately and observed the steady state values of five specific metabolites, strategically placed within the CCM: G6P and pyruvate (respectively entry and end points of glycolysis), R5P (middle of pentose phosphate pathway), palmitate (lipid production) and lactate (anaerobic energy production). The obtained values are illustrated in Fig. 6. They seem to confirm four main biological observations made in [6]:

- Glycolysis has a high activity in G1 (higher values of G6P* and PYR*),
- Lactate production is high in G1,
- Pentose phosphate increases in S and G2 (attested by higher values of R5P*),
- Lipid production increase in G2.

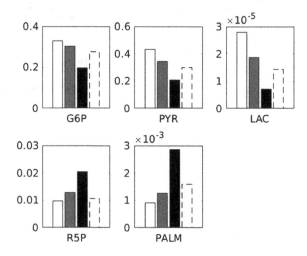

Fig. 6. Steady state concentrations of five key CCM metabolites in the G1 (white), S (grey) and G2 (black) models. For comparison, the dashed bars indicate the corresponding values for the original model.

4   Hybrid Simulation of Metabolism Along the Cell Cycle

Encouraged by the previous results, we decided to use a hybrid approach to observe the temporal succession of the three models, in the order imposed by biology. The idea was to reproduce, at least on a qualitative level, major metabolic variations as they are predicted theoretically or observed experimentally [3,5] during a whole cycle. In reality, the correct succession of cell cycle's phases is ensured by a complex regulatory network (see *e.g.* [7] for a discrete model in the case of mammalian cells), which is tightly linked to the cell's metabolism [9]. However, the interconnection remains difficult as the precise biochemical connections are not fully understood. Here, we decide to drastically simplify the "cell cycle" part by imposing a switch when the biomass reaches a certain level, thus mimicking the general growth of the mother cell before its division. Despite this drastic hypothesis, we show in the following that our approach is sufficient to reproduce expected metabolic variations during a cycle.

4.1   Creating a Hybrid Trajectory

We start by the description of the hybrid automaton. This automaton contains three modes corresponding to the three phases G1, S and G2. To each of these phases corresponds a model built in Sect. 3. In a given phase, we designate the state vector by $x^{phase}(t) = (x_I^{phase}(t), x_{II}^{phase}(t), x_{biomass}^{phase}(t))$. Starting in G1 at time t_0, we switch to S at time $t_1 > t_0$ and to G2 at time $t_2 > t_1$, where t_1 and t_2 are defined as follows:

- $x_{biomass}^{G1}(t_1) := x_{biomass} + \alpha x_{biomass} = (1 + \alpha)x_{biomass}^0,$
- $x_{biomass}^{S}(t_2) := x_{biomass} + \beta x_{biomass} = (1 + \beta)x_{biomass}^0,$

$x_{biomass}^0$ being the initial biomass. In other words, in this model the succession of phases is entirely determined by biomass evolution. α and β are parameters that verify $0 < \alpha < \beta < 1$ and correspond to fractions of biomass needed to be produced to enter the next phase. At each switch, we impose the continuity of the solution by setting:

$$x^{(i)}(t_i^+) = x^{(i-1)}(t_i^-).$$

To model the completion of a cycle at the end of the phase G2, we switch back to mode G1 at time $t_3 > t_2$ defined by $x_{biomass}^{G2}(t_3) := 2 \times x_{biomass}^0$. Thus, biomass has doubled and the cell can divide itself. We actually model an instantaneous mitosis by imposing:

$$x_{biomass}^{G1}(t_3^+) := \frac{x_{biomass}^{G2}(t_3^-)}{2}.$$

We present two hybrid trajectories in Fig. 7. As an initial condition, we use the stationary regime of the original model (see Fig. 2). By simulating the above hybrid system on several cycles, we observe that the trajectories ultimately stabilize to an oscillatory regime. The two hybrid trajectories depicted in Fig. 7 correspond to two set of values for parameters α and β. The choice of α and β seems to have an impact on a quantitative level, but not so much on a qualitative level. Overall, major metabolic variations through the succession of phases are retrieved [6]. First, there is a generally high glycolytic activity in G1, as illustrated by (generally) increasing G6P and pyruvate curves. Comparatively, pentose phosphate pathway and lipid production remain low. As expected, R5P starts to increase in S reproducing a high demand in nucleotide for DNA duplication. It is associated with a sharp fall of the ratio NADP/NADPH, confirming high activation of pentose phosphate pathway. This high level continues in G2, accentuated by a higher demand in lipid (palmitate).

Furthermore, the exchange metabolite curves allow a direct comparison with experimental curves given in [3]. Major effects are qualitatively recovered:

- a decrease of ATP during G1, followed by an increase during S and a decrease during G2,

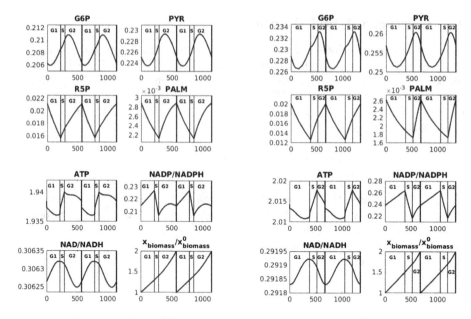

Fig. 7. Two examples of hybrid trajectories. Two cell cycles obtained for two different values of (α, β), left: $(0.3, 0.4)$, right: $(0.5, 0.7)$.

- a relatively low variation of the redox ratios[3],
- an increase of the NADP/NADPH ratio during G1, followed by a decrease during S and an increase during G2.

However, we observe a discrepancy in the NAD/NADH ratio: it decreases in G2 in the hybrid model whereas it seems to increase in experimental curves.

These first results seem promising: by combining the three versions of the model we were able to recover accurate biological observations made in [6] as well as experimental variations made in [3], thus validating the use of a hybrid approach to qualitatively capture the metabolic variations during the cycle. It is interesting to note that the modifications needed to build the hybrid system from the original ODE model are relatively parsimonious, which is always a good point from a modeling point of view. The obtained hybrid model is a good first step towards the analysis of the crosstalks between metabolism and cell cycle. If the cell cycle part is relatively simple (phases' succession is only imposed by a growth criterion), it is easy to imagine extensions including for instance a more complete representation of the cell cycle regulation network (as is [7]), or a more precise description of checkpoints as in [11]. In the next section, we propose a slight extension of the model to try to improve the model's predictions on a quantitative level.

---

[3] in [3], the range of variation of redox ratios is low in normal cells with respect to cancer cells.

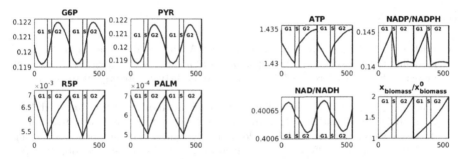

Fig. 8. Hybrid trajectory with updated palmitate demand $K_{PALM} = 2.55 \cdot 10^{-3}$. Parameters $(\alpha, \beta) = (0.3, 0.4)$.

4.2   Extension: Modifying Demand Parameters

If the hybrid simulations are relatively satisfying on a qualitative level, certain quantitative aspects still need improvement. In particular, from a dynamical system point of view the switch times seem a bit high (typically, a cell cycle is completed in tenth hours and not hundreds). By looking at the biomass ratio $\frac{x_{biomass}(t)}{x_{biomass}^0}$ in Fig. 7, we observed that its increase is particularly high in G2, when palmitate production is higher. We confirmed that by looking at the growth rate evolution $\mu(t)$ (not shown). Complementary simulations (not shown) also indicated that palmitate was usually the limiting factor of the growth rate. To alleviate this limitation, we decide to act on the demand of palmitate by reducing parameter K_{PALM} (see (4)). It was originally at $2.55 \cdot 10^{-2}$ and we reduced it to $2.55 \cdot 10^{-3}$. The results are illustrated in Fig. 8. Most of qualitative effects described previously are conserved, except for the ATP concentration during G2. The simulation time has been greatly reduced, indicating the demand parameters K_i in function μ are good candidates to calibrate the hybrid model.

5   Conclusion

The hybrid system proposed in this article is a preliminary work allowing us to analyse the temporal evolution of metabolites and fluxes through the phases of the cell cycle. First simulations are in good accordance with biological observations, at least at qualitative level. These positive results show that the hybrid approach is a promising way to combine metabolism and cell cycle. If the discrete part remains simplistic (the succession of phases is imposed), a next step will be to include more complete information about the cell cycle regulatory network [7,11] in order to better approach possible interconnections with the cell's metabolism.

Another extension concerns the incorporation of known disruptions in the regulations, thus proposing a hybrid model for cancer cells with dysfunctioning metabolism and cell cycle. Such a model, validated on experimental data such as in [3], would pave the way to find potential targets for therapy.

References

1. Barberis, M., Todd, R.G., van der Zee, L.: Advances and challenges in logical modeling of cell cycle regulation: perspective for multi-scale, integrative yeast cell models. FEMS Yeast Res. **17**(1), fow103 (2017). https://doi.org/10.1093/femsyr/fow103

2. Cai, L., Tu, B.P.: Driving the cell cycle through metabolism. Annu. Rev. Cell Dev. Biol. **28**(1), 59–87 (2012). https://doi.org/10.1146/annurev-cellbio-092910-154010

3. da Veiga Moreira, J., et al.: The redox status of cancer cells supports mechanisms behind the Warburg effect. Metabolites **6**(4), 33 (2016). https://doi.org/10.3390/metabo6040033

4. da Veiga Moreira, J., Hamraz, M., Abolhassani, M., Schwartz, L., Jolicœur, M., Peres, S.: Metabolic therapies inhibit tumor growth in vivo and in silico. Sci. Rep. **9**(1) (2019). https://doi.org/10.1038/s41598-019-39109-1

5. da Veiga Moreira, J., et al.: Cell cycle progression is regulated by intertwined redox oscillators. Theor. Biol. Med. Model. **12**(1) (2015). https://doi.org/10.1186/s12976-015-0005-2

6. Diaz-Moralli, S., Tarrado-Castellarnau, M., Miranda, A., Cascante, M.: Targeting cell cycle regulation in cancer therapy. Pharmacol. Ther. **138**(2), 255–271 (2013). https://doi.org/10.1016/j.pharmthera.2013.01.011

7. Fauré, A., Naldi, A., Chaouiya, C., Thieffry, D.: Dynamical analysis of a generic Boolean model for the control of the mammalian cell cycle. Bioinformatics **22**(14), e124–e131 (2006). https://doi.org/10.1093/bioinformatics/btl210

8. Funahashi, A., Matsuoka, Y., Jouraku, A., Morohashi, M., Kikuchi, N., Kitano, H.: Cell designer 3.5: a versatile modeling tool for biochemical networks. Proc. IEEE **96**(8), 1254–1265 (2008). https://doi.org/10.1109/JPROC.2008.925458

9. Kalucka, J., et al.: Metabolic control of the cell cycle. Cell Cycle **14**(21), 3379–3388 (2015). https://doi.org/10.1080/15384101.2015.1090068

10. Kaplon, J., van Dam, L., Peeper, D.: Two-way communication between the metabolic and cell cycle machineries: the molecular basis. Cell Cycle **14**(13), 2022–2032 (2015). https://doi.org/10.1080/15384101.2015.1044172

11. Novák, B., Tyson, J.J.: A model for restriction point control of the mammalian cell cycle. J. Theor. Biol. **230**(4), 563–579 (2004). https://doi.org/10.1016/j.jtbi.2004.04.039

12. Robitaille, J., Chen, J., Jolicoeur, M.: A Single dynamic metabolic model can describe mAb producing CHO cell batch and fed-batch cultures on different culture media. PLoS ONE **10**(9), e0136815 (2015). https://doi.org/10.1371/journal.pone.0136815

13. Segel, I.H.: Enzyme Kinetics: Behavior and Analysis of Rapid Equilibrium and Steady State Enzyme Systems. Wiley, New York (1993). Wiley classics library ed edn. oCLC: ocm31157638

14. Sheikh, K., Förster, J., Nielsen, L.K.: Modeling hybridoma cell metabolism using a generic genome-scale metabolic model of Mus musculus. Biotechnol. Prog. **21**(1), 112–121 (2008). https://doi.org/10.1021/bp0498138

15. Zhao, G., Chen, Y., Carey, L., Futcher, B.: Cyclin-dependent Kinase co-ordinates carbohydrate metabolism and cell cycle in S. cerevisiae. Mol. Cell **62**(4), 546–557 (2016). https://doi.org/10.1016/j.molcel.2016.04.026

Data-Informed Parameter Synthesis for Population Markov Chains

Matej Hajnal[2,4], Morgane Nouvian[1,3], David Šafránek[4], and Tatjana Petrov[2,3(✉)]

[1] Department of Biology, University of Konstanz, Konstanz, Germany
[2] Department of Computer and Information Science, University of Konstanz, Konstanz, Germany
tatjana.petrov@gmail.com
[3] Centre for the Advanced Study of Collective Behaviour, University of Konstanz, 78464 Konstanz, Germany
[4] Systems Biology Laboratory, Faculty of Informatics, Masaryk University, Botanická 68a, 602 00 Brno, Czech Republic

Abstract. Stochastic population models are widely used to model phenomena in different areas such as chemical kinetics or collective animal behaviour. Quantitative analysis of stochastic population models easily becomes challenging, due to the combinatorial propagation of dependencies across the population. The complexity becomes especially prominent when model's parameters are not known and available measurements are limited. In this paper, we illustrate this challenge in a concrete scenario: we assume a simple communication scheme among identical individuals, inspired by how social honeybees emit the alarm pheromone to protect the colony in case of danger. Together, n individuals induce a population Markov chain with n parameters. In addition, we assume to be able to experimentally observe the states only after the steady-state is reached. In order to obtain the parameters of the individual's behaviour, by utilising the data measurements for population, we combine two existing techniques. First, we use the tools for parameter synthesis for Markov chains with respect to temporal logic properties, and then we employ CEGAR-like reasoning to find the viable parameter space up to desired coverage. We report the performance on a number of synthetic data sets.

1 Introduction

Population models are widely used to model different phenomena: animal collectives such as social insects, flocking birds, schooling fish, or humans within

TP's research is supported by the Ministry of Science, Research and the Arts of the state of Baden-Württemberg, and the DFG Centre of Excellence 2117 'Centre for the Advanced Study of Collective Behaviour' (ID: 422037984), MH's research is supported by Young Scholar Fund (YSF), project no. $P83943018FP430\_/18$. MN's research is supported by the Mentorship grant from the Zukunftskolleg. DŠ's research is supported by the Czech Grant Agency grant no. GA18-00178S.

© Springer Nature Switzerland AG 2019
M. Češka and N. Paoletti (Eds.): HSB 2019, LNBI 11705, pp. 147–164, 2019.
https://doi.org/10.1007/978-3-030-28042-0_10

societies, as well as molecular species inside a cell, cells forming a tissue. Animal collectives show remarkable self-organisation towards emergent behaviours without centralised control. Quantitative models of the underlying mechanisms can directly serve important societal concerns (for example, prediction of seismic activity [27]), inspire the design of distributed algorithms (for example, ant colony algorithm [17]), or aid robust design and engineering of collective, adaptive systems under given functionality and resources, which is recently gaining attention in vision of smart cities [22,26]. Quantitative prediction of the behaviour of a population of agents over time and space, each having several behavioural modes, results in a high-dimensional, non-linear, and stochastic system [20]. Hence, computational modelling with population models is challenging, especially when the model parameters are unknown and experiments are expensive.

In this paper, we investigate how to obtain the parameters for single agent behaviour, based on data collected for a population. Measurements for different population sizes are especially important when studying social feedback: an adaptation of individual's behaviour to the changing context of the population. For example, honeybees protect their colonies against vertebrates by releasing an alarm pheromone to recruit a large number of defenders into a massive stinging response [28]. However, these workers will then die from abdominal damage caused by the sting tearing loose [33]. In order to achieve a balanced trade-off towards efficient defence, yet no critical worker loss, each bee's response to the same amount of pheromone may vary greatly, depending on its social context, which, in the case of bees, has been experimentally validated.

To tackle this problem, we assume a simple communication scheme among identical individuals, such that n individuals together form a discrete-time Markov chain (DTMC) M with at most n parameters. Each population eventually reaches one of its terminal strongly connected components (tSCC) in the underlying MC. We employ the theoretical steady-state assumption that is commonly used in biological modelling scenarios: we assume that the experimental observations can be taken when the steady state is reached, hence that experimental measurements allow us to estimate probabilities of reaching any of the tSCCs in the form of a confidence interval (for any desired confidence level α). We assume $\mathcal{V}$ denotes a set of model parameters, each defined over domain $[0, 1]$. Our major goal is to synthesise a *viable parameter space* Θ, $\Theta \subseteq [0, 1]^{|\mathcal{V}|}$, such that the following condition is satisfied:

$$\theta \in \Theta \text{ if and only if } M(\theta) \models \bigwedge_{\text{all } tSCCs} \varphi(tSCC \mid \texttt{data}) \tag{1}$$

where $\varphi(tSCC \mid \texttt{data})$ expresses that reaching a tSCC is achieved within the confidence interval estimated from experimental data. In contrast to traditional parameter inference techniques which return a single estimate, the parameter synthesis approach gives a quantitative characterisation of the entire domain of satisfying parameter values.

We propose and implement a workflow for obtaining the viable parameter space for a simple population model. The workflow, summarised in Fig. 3 (concrete steps and used technology), and Algorithm 1 (region generation and splitting), consists of two steps: first, we obtain a symbolic characterisation of the distribution over tSCCs in form of multivariate rational functions, leveraging the existing tools for parameter synthesis. Second, we employ CEGAR-like reasoning (candidate region generation and checking) for determining the viable parameter space, until the desired proportion of the domain (called coverage) is reached. We implemented several variants of the search algorithm, and tested the performance on synthetic data sets. It is worth noting that, for showcasing the framework, we here implemented a workflow for a specific class of population models which have acyclic underlying transition system and each tSCC contains one state (being strongly inspired by the propagation of alarm pheromones in honeybees). The framework is applicable to any general DTMC and any persistence or repeated reachability temporal logic property.

Related Work. Population models induced from a counting abstraction, have been widely studied in the context of biochemical reaction networks. Ideas focusing on faster prediction of resulting distributions over sub-populations of molecular species, based on fluid, continuous space approximations, [6,7], as well as moment closure approximations [3,4,21] could be useful for improving scalability of our parameter synthesis problem. Further promising approaches include global optimisation algorithms adopted from machine learning ideas, allowing to develop a notion of robustness degree in [5,8]. Different to our work here, these approaches handle continuous-time Markov models, and general temporal specifications. Population protocols – interacting identical agents, each being a state machine – have been extensively studied in the context of distributed algorithms [1,2], where the focus is different than ours: it is on the performance of implementation of protocols such as, for example, leader election or self-stabilisation.

When it comes to exact methods for parameter synthesis, there exists a substantial body of work on verification of parametric discrete-time Markov chains (pMC), subject to temporal logic properties: symbolic computation of reachability properties through state elimination in a parametric Markov chain [13,23], lifting the parameters towards verifying a non-parametric Markov decision process (MDP) instead of the original pMC [31], candidate region generation and checking, helped by SMT solvers (see [24] and references therein). Specifying biological properties as temporal logic formulae, and using such specifications for parameter synthesis, has already been applied in biological modelling: in [19], the authors compute the robustness of evolving gene regulatory networks by first synthesising the viable space of parameters with constraint solvers. In a related setup in [10,11,36], the authors express properties of general biochemical reaction networks in continuous signalling logic (CSL), where they deal with the parameter synthesis for continuous-time Markov chains. Only recently, in [30], direct integration of data into Bayesian verification of parametric chains has been proposed. To the best of our knowledge, the latter framework could not directly handle our case study, because it is designed to handle affine transition functions in the pMC.

In our computational experiments, we considered using several tools which support parameter synthesis - PRISM [25], Prophesy [15], and Storm [16]. Finally, we used PRISM as it supports a command line input, helpful for the automatisation of the workflow.

2   Model

Consider n identical agents, each given a task. Each agent succeeds at completing a task with probability p, and fails with probability $(1-p)$. If there is no other interaction among agents, the number of successes is binomially distributed. We add the following type of communication: an agent who fails has a second chance to succeed if some of the other agents already succeeded. In other words - any agent that succeeds can help other agents who have failed in the first attempt[1]. We assume that each agent has two chances to succeed, but not more (allowing unbounded number of chances would lead to an uninteresting case where either all agents fail immediately or, otherwise, all agents succeed eventually). In Fig. 1, we sketch the states of a single agent and the transitions between them. We consider two model variants: the probability of conversion in the second attempt is constant (Fig. 1, middle), or it can vary depending on the number of agents who have already succeeded (Fig. 1, right).

2.1   Preliminaries

Let $M = (S, P, \xi_{in})$ be a transition system inducing a discrete-time Markov process (DTMC) $\{X_t\}$ over finite state space S, with kernel P, initial distribution ξ_{in}. We denote by P_M the probability measure over the respective prefix sets of traces, defined in a standard way: inductively by $\mathsf{P}_M(s) = \xi_{in}(s)$ and $\mathsf{P}_M(\overline{s_0 \ldots s_k s}) = \mathsf{P}_M(s_0 \ldots s_k)P(s_k, s)$, where $s_0, \ldots, s_k, s \in S$.

A population of agents that we assume in our case study will follow a discrete-time Markov chain. In order to define logical properties over it, we add labels to the states in a standard way.

Definition 1. *Labelled Markov Chain (LMC)* is a tuple $M = (S, P, \xi_{in}, AP, L)$ over finite state space S, transition probability function $P : S \times S \to [0, 1]$ such that $\sum_{s' \in S} P(s, s') = 1$ for all $s \in S$, initial distribution $\xi_{in} : S \to [0, 1]$, set of atomic propositions AP, and a state-labelling function $L : S \to 2^{AP}$.

A finite run of LMC $M = (S, P, \xi_{in}, AP, L)$, denoted by $\sigma = (s_0, s_1, \ldots) \in S^l$, induces a trace $\tau(\sigma) \in (2^{AP})^l$, defined inductively by $\tau(s) = L(s)$ and $\tau(\overline{\sigma s}) = \tau(\sigma)L(s)$. The prefix set of traces defined by τ inherits the probability measure P_M from the DTMC (S, P, ξ_{in}), hence $\mathsf{P}_M(\sigma)$ denotes the probability of a prefix

---

[1] In our case, 'help' does not involve interaction between agents, - it is simultaneously broadcasted from an agent to all the others.

set of traces σ. Consider a fragment of linear temporal logic (LTL) properties defined over the traces for LMC M, induced by the grammar $\phi := \varphi \in AP ::$ $FG(\phi)$. We will write $P(M \models \phi) = P_M(\{\tau \in \Sigma^\omega \mid \tau \models \phi\})$ to denote the probability of satisfaction of LTL property ϕ in the LMC M.

When the transition probabilities are not known, but rather are rational functions of some parameters from the parameter set $\mathcal{V}$, each over domain $[0, 1]$, we work with a parametric Labelled Markov Chain (pMC). We here restrict our attention to the case when the transition probabilities are multivariate polynomial functions over the variables $\mathcal{V}$, which we will denote by Poly.[2]

Definition 2. *Parametric Markov Chain (pMC) is a tuple* $M_\mathcal{V} = (S, P_\mathcal{V}, \xi_{in},$ $AP, L, \mathcal{V})$, *where* $P_\mathcal{V} : S \times S \to \mathsf{Poly}$ *defines the probability transition matrix, and for each evaluation of parameters* $\theta \in [0, 1]^{|\mathcal{V}|}$ *induces a Markov chain* $M(\theta) = (S, P_\theta, \xi_{in}, AP, L)$, *where* P_θ *denotes the instantiation of the expression* $P_\mathcal{V}$, *for parameter evaluations given by a vector* θ. *Consequently, for all* $\theta \in [0, 1]^{|\mathcal{V}|}$, *for all* $s \in S$, $\sum_{s' \in S} P_\theta(s, s') = 1$.

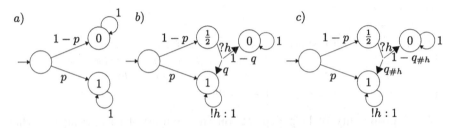

Fig. 1. Single-agent model. (a) From the initial state, the success (state labelled with **1**) is realised with probability p. (b) If no success is achieved in the first attempt (state $\frac{1}{2}$), and if another agent helps (input $?h$), the transition to a success state occurs with probability q. Otherwise, the transition to a final state occurs. In case of success, the agent keeps emitting a help action (output $!h$). (c) Probability of success after being helped depends on the number of successes in the population.

In the next subsection, we define the behaviour of a population of n agents, as a pMC. Denote by $[x, y]$ the range $\{x, x + 1, \ldots, y\}$, when it is clear from the context that x and y are integers.

2.2 Case Study

Let $n \in \mathbb{N}_{>0}$. Interactions among n identical agents can be captured by a pMC $M_\mathcal{V}^{(n)} = (S, P_\mathcal{V}, s_0, L, AP, \mathcal{V})$, where $S = \{s_0, s_1, \ldots\}$ is a finite set of states, and the set of variables $\mathcal{V} = \{p, q_1 \ldots, q_{n-1}\}$. We assign atomic propositions to reflect

---

[2] In general, the reachability probabilities for a pMC can be expressed by rational functions; In our case study, polynomials will suffice because the underlying transition system is acyclic.

the subpopulations of agents who succeeded, failed, or have another chance, denoted by $\mathbf{1}$, $\frac{1}{2}$ and $\mathbf{0}$, respectively: $AP = \{0, 1, \ldots, n\}^{\{\mathbf{1}, \frac{1}{2}, \mathbf{0}\}} \cup \{\text{INIT}\}$. The labelling function counts the number of success outcomes ($\mathbf{1}$), failed outcomes ($\mathbf{0}$), and those with another chance ($\frac{1}{2}$).

Initial state has label INIT. After one step, the population can be in one of $(n + 1)$ different states, counting $k \in [0, n]$ successes and $(n - k) \in [0, n]$ agents in state $\frac{1}{2}$. Since the initial probability of each success is p, the probability to move to a state with k successes in the first step is $\binom{n}{k} p^k (1 - p)^{n-k}$. Hence, after the first step, the number of successes follows a binomial with parameter p. Notice that there is a possibility of having zero successes after the first step – state labelled with $(\frac{1}{2}, \ldots, \frac{1}{2})$ – which is already a tSCC, since no help will be possible for any of the agents. In the second step, in case there are $k > 0$ agents who can help, each of the agents with label $\frac{1}{2}$ may be helped with probability q_k, and it will never succeed (state labelled with $\mathbf{0}$) with probability $(1 - q_k)$. If all agents who will be helped, do so simultaneously, then the probability that c out of $n - k$ agents in state $\frac{1}{2}$ will succeed is $\binom{k}{c} q_k^c (1 - q_k)^{k-c}$, inducing the following transition function $P_{\mathcal{V}} : S \times S \to \text{Pol}(\mathcal{V})$:

(synchronous semantics)

$P_{\mathcal{V}}^{syn}(s, s') :=$

$$
\begin{cases}
\binom{n}{k}(1-p)^k p^{n-k} & \text{if } L(s) = \text{INIT}, L(s') = (k, n - k, 0) \text{ for some } k \in [0, n] \\
\binom{k}{c}(1-q_k)^c q_k^{k-c} & L(s) = (k, n - k, 0), L(s') = (k + c, 0, n - k - c) \text{ for some } c \in [0, k], \\
1 & \text{if } L(s) = (k, 0, n - k) \text{ and } L(s') = (k, 0, n - k) \\
& \text{or } L(s) = L(s') = (0, n, 0).
\end{cases}
$$

When probability of help depends on the number of success agents (that is, $q_1, \ldots q_{n-1}$ seen as a function is not constant), it is unrealistic to assume simultaneous success of all agents who will be helped: as soon as one of the agents with label $\frac{1}{2}$ succeeds, the probability of success for other agents with state $\frac{1}{2}$ changes. Accounting for such non-synchronous update of the population state defines a different transition function $P_{\mathcal{V}} : S \times S \to \text{Pol}(\mathcal{V})$ in our population pMC:

(semi-synchronous semantics)

$P_{\mathcal{V}}^{semisyn}(s, s') :=$

$$
\begin{cases}
\binom{n}{k}(1-p)^k p^{n-k} & \text{if } L(s) = \text{INIT}, L(s') = (k, n - k, 0) \text{ for some } k \in [0, n] \\
q_k & L(s) = (k, n - k - c, c), L(s') = (k + 1, n - k - c - 1, c), \\
1 - q_k & L(s) = (k, n - k - c, c), L(s') = (k, n - k - 1 - c, c + 1), \\
1 & \text{if } L(s) = (k, 0, n - k) \text{ and } L(s') = (k, 0, n - k) \\
& \text{or } L(s) = L(s') = (0, n, 0).
\end{cases}
$$

where $c \in [0, k]$.

In Fig. 2a, we illustrate a population pMC for $n = 3$, with semi-synchronous update. In general, $P_{\mathcal{V}}(\cdot, \cdot)$ are multivariate polynomials of maximum order n. Population model with n agents will have $n + 1$ terminal states (tSCCs) with respectively $0, 1, \ldots, n$ success outcomes. In case of three agents, the semi-synchronous population model counts eight states (Fig. 2), and in general, the state space counts $O(n^2)$ states; In the case of a synchronous update, the number of states is $O(n)$.

Remark. A population of n agents can be formally defined as a parallel composition of n agents, with variants depending on how the synchronisation on help actions is defined [34,35,37]. Then, the counting abstraction is a lumpable partition of the state space underlying the parallel composition model. Deriving formally the parallel composition of generic communicating agents is beyond the scope of this manuscript, so we directly define the population quotient.

3   Parameter Synthesis

Our general goal is to synthesise the parameters of single-agent behaviour, based on data collected for a population. More concretely, we assume that we can observe the system after the underlying population Markov chain has reached one of its tSCCs.

3.1   Symbolic Expressions for Measured Properties

In Fig. 2b, we show the distribution among the tSCCs as polynomial expressions over parameters. In general,

$$\text{let } F^{(n)}(k \mid \mathcal{V}) \in \mathsf{Pol}^{(n)}(\mathcal{V}) \text{ be such that} \tag{2}$$

$$\text{for all } \theta \in [0,1]^{|\mathcal{V}|}, \;\; F^{(n)}(k \mid \theta) = \mathsf{P}\left(M^{(n)}(\theta) \models FG(k, 0, n-k)\right), \tag{3}$$

that is, $F^{(n)}(k \mid \mathcal{V})$ is a polynomial expression over variables $\mathcal{V}$, exactly characterising the reachability of a tSCC counting k successes in a population pMC $M_{\mathcal{V}}^{(n)}$. We omit superscript (n) and subscript $\mathcal{V}$ when clear from context. Notice that the formula does not involve the information obtained from data - it refers to the probability of eventually reaching a tSCC with k successes, as a function of parameters. In the implementation, we will leverage existing model checking tools to obtain these polynomials.

3.2   Data

For a population of size n, we assume N experiments in which we can observe the number of successes at steady state (for the purpose of performance analysis reported in Sect. 4, we will obtain synthetic data, by simulating the experiments on a computer). Denote by $X_i \in [0,n]$ the outcome in experiment $i = 1, \ldots, N$. As we have proportional data, we can estimate the probability of reaching each of the tSCCs in a standard way:

for given confidence level α, for all $k \in \{0, 1, 2, \ldots, n\}$,

$$\textsf{data}: \mathsf{P}_{M(\mathcal{V})}(FG(k, 0, n-k)) \in \hat{x}_k \pm \left(z_{\alpha/2}\sqrt{\frac{\hat{x}_k(1 - \hat{x}_k)}{N}} + \frac{0.5}{N} \right), \tag{4}$$

where $\hat{x}_k := \frac{X_k}{N}$ is the point estimate for the probability of eventually having k success agents in a population of n agents. The terms $\frac{0.5}{N}$ are included for the correction of approximation from discrete to continuous distribution. In the experiments, we choose sample size $N = 100$ and $N = 1500$, that guarantee margins to be at most 0.1 and 0.025 respectively. In further text, we will denote by $[\hat{x}_l, \hat{x}_u](k \mid \text{data})$ the lower and upper bound of the confidence interval for tSCC with k successes.

a)

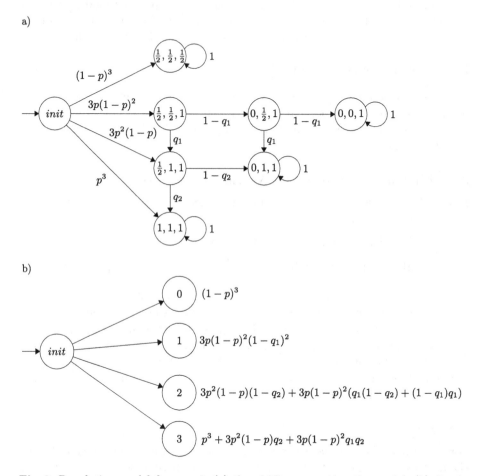

b)

Fig. 2. Population model for $n = 3$: (a) the pMC representing the model, (b) distribution among possible final states is a list of $2n$-degree multivariate polynomial over model parameters $\mathcal{V} = \{p, q_1, \ldots, q_{n-1}\}$.

3.3   Region Generation and Refinement with Constraint Solvers

Back at the general question introduced in Eq. (1), the conjunction of properties $\varphi(tSCC \mid \text{data})$ for our case study amounts to:

$$\bigwedge_{k=0}^{n} \left(F^{(n)}(k \mid \mathcal{V}) \in [\hat{x}_l, \hat{x}_u](k \mid \mathtt{data}) \right), \tag{5}$$

expressing that each of the tSCCs is reached with probability within a confidence interval obtained from data.

Every parameter evaluation $\theta \in [0,1]^{|\mathcal{V}|}$ such that the above constraint holds, belongs to our goal viable set Θ, and vice versa. A single point estimate may be satisfactory in some cases. However, for smaller data samples, we wish to have a better idea of the viable parameter space, and we wish to explore the parameter space further. We can pass a query $\exists \theta \in (\mathcal{V} \mapsto [0,1]^{|\mathcal{V}|})$, such that $\bigwedge_{k=0}^{n} F^{(n)}(k \mid \mathcal{V}) \in [\hat{x}_l, \hat{x}_u](k \mid \mathtt{data})$ to an SMT solver, such as Z3 [14]. Then, depending on the outcome, we further refine the parameter space in CEGAR-like fashion into

- "safe" or "green" regions, where the constraints are met,
- "unsafe" or "red" regions, where the constraints are not met,
- "unknown" or "white" regions, where the constraints may hold or not,

the idea of which is also used by the existing tools, such as Prophesy [15]. For each parameter evaluation in a safe region, the formula holds because the negation of the formula is not satisfiable. For each parameter evaluation in the unsafe region, the formula itself is not satisfiable. The unknown region is not yet refined, since it may contain both – parameters for which formula holds and for which it does not hold. In our implementation, we use a naive refinement approach, splitting into two halves along dimension with the largest range. This split occurs when the given region is proven to be neither safe nor unsafe. As the main stopping criterion, we introduce the parameter $\mathtt{coverage}$, such that the fraction of the explored parameter space and the whole parameter space is larger than coverage:

$$\Theta_{green} + \Theta_{red} > \mathtt{coverage}.$$

Algorithm. Algorithm 1 summarises the idea of this procedure. Depending on how the query in Eq. (5) is passed to the solver, we used three major ideas:

- *All-in-one approach* (described in Algorithm 1). Each clause in the conjunction in Eq. (5), is joined using logical operators. This may result in computation overhead because in each splitting, the solver has to first find a point which satisfies each of the properties, and then prove the unsatisfiability of its negation.
- *Prop-by-prop approach.* Each clause in the conjunction in Eq. (5) is queried separately, and the resulting regions are conjuncted in the end.
- *Iterative approach.* In this approach, clauses are also passed to the solver separately, and each next clause is not resolved for the whole domain of parametrisation, but only for the portion of the parametrisation that are found not be in a red region so far.

All three approaches can be parallelised. The easiest to parallelise is the second approach, since each clause is checked independently. In case of iterative method, sorting properties by amount of unsafe region may help to outperform the other settings.

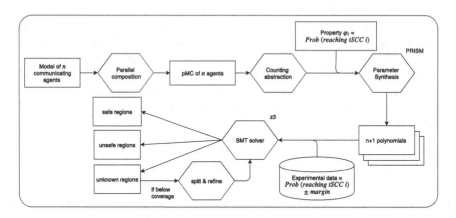

Fig. 3. Workflow used to analyse the case study models, where model pre-processing, composition and the counting abstraction (lumpable quotient) are computed in one step. In our case study we used only reachability properties. The output of parameter synthesis are rational functions.

4   Implementation and Results

We implemented the workflow in a Jupyter notebook [18]. First, two-parametric and multiparametric models ($\mathcal{V} = \{p, q\}$ and $\mathcal{V} = \{p, q_1, \ldots, q_{n-1}\}$) are created with respective property files, for desired population size. We implemented the construction for three kinds of compositions – *synchronous*, *asynchronous*, and *semisynchronous*. Asynchronous models are largest and provide multi-affine labelling, hence are usable for Storm model checker. *Semisynchronous models* are instantiated in Fig. 2. Polynomials $F^{(n)}$ are obtained with PRISM [15], by invoking parametric model-checking on properties $FG(k, 0, n - k)$, for $k \in [0, n]$. The implementation includes sampling and visualisation of polynomials $F^{(n)}\theta(k \mid \theta)$ for given parametrisation $\theta \in [0, 1]^{|\mathcal{V}|}$. We generated two synthetic data sets by first generating the parametrizations $\theta_1, \theta_2 \in [0, 1]^{|\mathcal{V}|}$ uniformly at random, and then simulating the experimental outcomes $X_1, X_2, \ldots, X_N$. Finally, we employed an SMT solver Z3 [14] to obtain the viable parameter space by querying on the constrains specified in Eq. (5). The resulting partitioning of the parameter domain can be visualised as green and red regions (Fig. 4).

It is possible to directly search for the viable parameter space, without first generating polynomials $F^{(n)}(k)$, by non-parametric model checking for the constraints in Eq. (6):

$$\Theta = \{\theta \in [0,1]^{|\mathcal{V}|} \mid$$

$$\bigwedge_{k=0}^{n} \mathsf{P}_{\geq \hat{x}_l}\left(M(\theta) \models FG(k, 0, n-k)\right) \wedge \mathsf{P}_{<\hat{x}_u}\left(M(\theta) \models FG(k, 0, n-k)\right)\}$$

(6)

In order to compare the scalability of our approach with the existing model checkers, we also implemented a search for Θ, as non-parametric model checking for the constraints in Eq. (6) in PRISM and Storm [16].

The implementation is written in Python, calling model-checking tool PRISM and Z3. For Storm, commands usable for data-driven parameter synthesis of the case study are provided within the notebook. Created source code with generated data is provided within the Jupyter notebook which is available at www.fi.muni. cz/~xhajnal/hsb2019/. A scheme of the workflow is visualised in Fig. 3.

4.1   Performance

The first bottleneck of the workflow was parameter synthesis procedure due to insufficient memory and exception of Java code when running PRISM. More precisely, using default setting we have obtained results for 35 agents using two_param model and for 15 agents in multi_param models. Increasing Java heapsize (from default 1 GB) to 8 GB we obtained results for 50 agents using two_param model, but did not get results for 18 agents in multi_param models due to memory. This problem was solved by analysing each clause in Eq. (5) separately (*prop-by-prop approach*, explained in Sect. 3.3). We were able to compute results in one hour for 75 agents using two_param and even for 18 agents in multi_param models. The results are summarised in Table 1.

The second bottleneck of the workflow was the region generation and refinement using constraint solvers. Using naive algorithm splitting the parameter space in each dimension recursively calling itself, we have obtained data for population up to five agents within a minute. This approach was not sufficient for greater populations or multi-parametric models, hence we do not compare it with further options.

The first improvement was refining only one dimension at a time – *Alg1*. This version is represented in Algorithm 1. Problems occurred with increasing coverage_threshold and max_recursion_depth. It is because a preorder traversal provided by the simple recursion searches smaller and smaller regions in each iteration, while the second half of the region, possibly green or red, will be checked only after the whole first half is split[3]. To improve this issue we added a queue to provide lever-order traverse – *Alg2*.

Further, we exploit the information as to which satisfiable subregion contains the found model. With this information, some checks for unsatisfiability can be skipped. This change is implemented in *Alg3*. The improvement of computation

---

[3] If the coverage is not set below 50%.

Table 1. Parameter synthesis using PRISM for respective model and properties (reaching a tSCC). When memory limit, 8 GB java_heapsize, exceeded (marked as OM) used prop-by-prop approach.

model_type, #agents	#params	#states	#transitions	param_synthesis
semisynchronous_2	2	9	12	2 s
semisynchronous_3	2	13	19	2 s
semisynchronous_5	2	24	39	2 s
semisynchronous_10	2	69	124	2 s
semisynchronous_20	2	234	444	4 s
semisynchronous_50	2	1329	2604	6,5 m
semisynchronous_75	2	2929	5779	OM/1 h
semisynchronous_3	3	13	19	2 s
semisynchronous_5	5	24	39	2 s
semisynchronous_10	10	69	124	2 s
semisynchronous_15	15	139	259	15 s
semisynchronous_18	18	193	364	OM/2,5 m
semisynchronous_20	20	234	444	OM/OM

time for respective algorithm is shown in Appendix Table 2. Since the computation time depends heavily on the intervals constraining the polynomials, to conclude results on scalability one would need to run more cases.

After these improvements, we were able to obtain results for 10 agents for two_param models. In the experiments, we choose sample size $N = 100$ and $N = 1500$, that guarantee margins to be at most 0.1 and 0.025 respectively. The results are visualised in Fig. 4. With increasing sample size, the unsafe regions grow much quicker than with increasing the population number.

Finally, we compare the computation time of *Alg1-3* with PRISM and Storm results in Appendix Table 3. All results were obtained using Freya - desktop, i7-8700K, GeForce GTX 1060, 32 GB RAM, with SSD disk.

To view, run, or edit the case study use www.fi.muni.cz/~xhajnal/hsb2019/. Html files serve to view analysis without using Jupyter, the zip file contains notebooks and Data sets.

5   Discussion and Future Work

In this paper, we proposed how to synthesise parameters for a class of population discrete-time Markov Chains, for which we assume that experimental data at steady state is available. We illustrated the workflow on a case study over n identical, interacting agents with n parameters, subject to persistence properties. Our approach is exact, in the sense that the only approximation involved is the estimation of confidence intervals when the property is extracted from the data. Moreover, our parameter search is agnostic in the sense that we make no assumption on the dependency between parameters $q_1, q_2, \ldots, q_{n-1}$. As a result,

---

Algorithm 1. Space Refinement

---

1 **Function** check([*interval*], [*property*], [*data*]):
2   **if** *sat(∀parameter.value() ∈ interval ∧ ∀property ∈ (data ± margin))* **then**
3   **if** *check_safe([interval], [property], [data]) == safe* **then**
4   **return** safe
5   **else**
6   **return** model
7   **else**
8   **return** unsafe

9
10 **Function** check_safe([*interval*], [*property*], [*data*]):
11   **if** *sat(∀parameter.value() ∈ interval ∧ ∃property ∉ (data ± margin))* **then**
12   **return** unsat
13   **else**
14   **return** safe

15 **Function** check_deeper([*interval*], [*property*], [*data*], *recursion_depth, min_rectangle, coverage_threshold*):
16   **if** *(recursion_depth == 0) ∨ ([interval].size() < min_rectangle) ∨ (coverage() > coverage_threshold)* **then**
17   **return**
18   **if** *check([interval], [property], [data]) == safe* **then**
19   *green.add([interval])*
20   **else if** *check([interval], [property], [data]) == unsafe* **then**
21   *red.add([interval])*
22   **else**
23   check_deeper([*interval*].left(), [*property*], [*data*], *recursion_depth* − 1, *min_rectangle, coverage_threshold*)
24   check_deeper([*interval*].right(), [*property*], [*data*], *recursion_depth* − 1, *min_rectangle, coverage_threshold*)

25 **Function** setup([*interval*], [*property*], [*data*], *recursion_depth, min_rectangle, coverage_threshold*):
26   *whole_area = [interval].size()*
27   *green = {}*
28   *red = {}*
29   *coverage() = union(green, red)*
30   check_deeper([*interval*], [*property*], [*data*], *recursion_depth, min_rectangle, coverage_threshold*)
31   **return** (green, red)

---

we may be able to discover unexpected dependencies in the viable region, but this comes at the cost of the number of parameters growing directly with the population size. Synthetic data set allowed us to identify several scalability bottlenecks

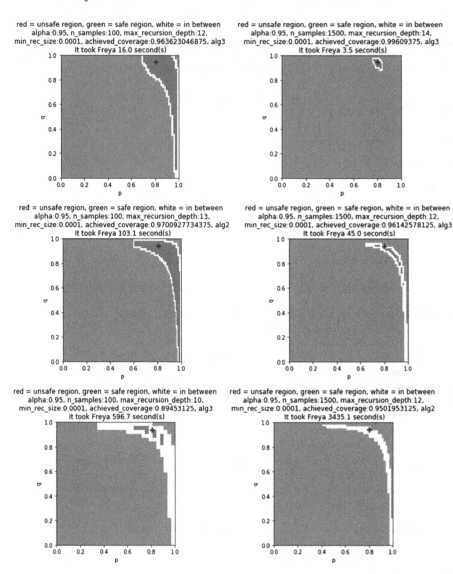

Fig. 4. Visualisation of results of *Alg2, 3* for two-param models with populations of 2, 5, and 10 agents in the respective row, and sample size of 100 and 1500 in the respective column with constant value of 0.005 absolute margin added to data. Parameter point from which the data were obtained is $p = 0.81, q = 0.92$ - shown as a blue cross. (Color figure online)

and open directions for future research. We implemented three algorithms refining the parameter space using a SMT solver: *Alg1* – simple recursion splitting region into two in dimension with largest interval, *Alg2* – level order traversal instead of preorder, *Alg3* – passing information in which subregion the model was found. These changes improved the performance – see Appendix Table 2.

We plan to further investigate how to improve the scalability towards larger population size and more parameters, especially in light of the fact that the realistic scenarios involve more complex communication schemes and hundreds or thousands of individuals. There are several ideas we plan to consider for improving the scalability. One idea includes bypassing the time-expensive calls to the constraint solver by other algorithms, such as worst-case interval propagation through a Markov decision process (MDP) [31], as well as different encoding of measurements: encoding each outcome as separate property is impractical when the sample size is small, or when only two parameters define the process. Further immediate ideas include moment closure approximations [3,4,21], pre-sampling of the parameter space in order to quickly label large portions or red or green areas, as well as employing Bayesian inference. The latter has been extensively used for parameter inference over continuous-time Markov chains of biochemical reaction networks [9,32]. Finally, the generalisation of the class of properties to handle systems with multi-state tSCCs would require extracting properties from data, for models where tSCCs contain more than one state, which in turn requires a detection of tSCC states in a black-box setting. To this end, we will consider the statistical model checking approach proposed in the recent works [12,29], where the tSCCs are detected with desired confidence, subject to only an assumption that each transition probability exceeds a threshold. On a different front, we plan to explore formal properties suitable for describing time-series measurements, which will require a Markov chain model in continuous time.

A Performance Comparison

Table 2. Runtime comparison for parameter space refinement using our algorithm – *Alg1-3*. Computation times for respective population size, two/multi param case, data, algorithm, and setting. All models used in the comparison are semisynchronous and the results were computed using *all-in-one* approach. Computation time in seconds.

		two_param								multi_param	
#agents		2		3		5		10		3	
data_set		1	2	1	2	1	2	1	2	1	2
alpha, n_samples		cov_thresh = 0.95, recursion_depth = 15									
0.95;**100**	*Alg1*	37.9	19.1	56.5	25.1	133	TO	TO	TO	48.9	29
	Alg2	8	2.3	13.1	2.3	44.6	TO	TO	TO	0.9	0.56
	Alg3	6.8	1.7	11.1	2.1	41.2	TO	TO	TO	0.7	0.5
0.95;**1500**	*Alg1*	0.2	6	0.3	8.1	0.1	30.8	TO	0.3	6	29.9
	Alg2	0.2	0.3	0.3	0.4	0.1	0.6	346.3	0.3	0.6	0.7
	Alg3	0.2	0.2	0.3	0.3	0.1	0.5	307	0.3	0.5	0.6
0.95;**3500**	*Alg1*	0.2	3.6	<0.1	5.4	0.1	17.2	3.9	0.3	3.9	3
	Alg2	0.2	0.2	<0.1	0.3	0.1	0.5	3.7	0.3	0.5	0.3
	Alg3	0.2	0.2	<0.1	0.3	0.1	0.5	3.7	0.3	0.5	0.3

Table 3. Runtime comparison of our algorithm, PRISM, and Storm. Our algorithm, *Alg3*, and PRISM were using semisynchronous models and Storm used asynchronous models. Rows for *Alg3* contain the time used to compute polynomials $F(k)$ (done by PRISM), plus the time needed to refine the space. By N/A we denote cases when Storm was unable to return a result due to a technical problem. We used Data set 1.

	two_param				multi_param		
#agents	2	3	5	10	3	5	10
alpha, n_samples, approach	0.95; **100**; *all-in-one*						
PRISM	2.5 s	4.5 s	7.8 s	30 s	5 m50 s	TO	TO
Alg3	2 + 6.8 s	2 + 11 s	2 + 41 s	TO	2 + 0.7 s	TO	TO
alpha, n_samples, approach	0.95; 100; *thresh-by-tresh*						
PRISM	5 s	7 s	12 s	33 s	3 m51 s	TO	TO
Storm	<0.1 s	<0.1 s	1 s	2 s	6 s	N/A	N/A
alpha, n_samples, approach	0.95; **1500**; *all-in-one*						
PRISM	3.7 s	5 s	7.8 s	37.5 s	8 m	TO	TO
Alg3	2 + 0.2 s	2 + 0.3 s	2 + 0.1 s	2 s + 5.1 m	2 + 0.5 s	TO	TO
alpha, n_samples, approach	0.95; 1500; *thresh-by-tresh*						
PRISM	9 s	9 s	12 s	44 s	6 m	TO	TO
Storm	0.00 s	1 s	1 s	2 s	8 s	N/A	N/A
alpha, n_samples, approach	0.95; **3500**; *all-in-one*						
PRISM	4 s	5.2 s	10 s	40 s	9 m	TO	TO
Alg3	2 + 0.9 s	2 + 0.0 s	2 + 0.1 s	2 + 0.1 s	2 + 0.5 s	TO	TO
alpha, n_samples, approach	0.95; 3500; *thresh-by-tresh*						
PRISM	8 s	10 s	11 s	45 s	6 m51 s	TO	TO
Storm	<1 s	1 s	1 s	3 s	8 s	40 m	N/A

References

1. Alistarh, D., Gelashvili, R., Vojnović, M.: Fast and exact majority in population protocols. In: Proceedings of the 2015 ACM Symposium on Principles of Distributed Computing, pp. 47–56. ACM (2015)
2. Aspnes, J., Ruppert, E.: An introduction to population protocols. In: Garbinato, B., Miranda, H., Rodrigues, L. (eds.) Middleware for Network Eccentric and Mobile Applications, pp. 97–120. Springer, Heidelberg (2009). https://doi.org/10.1007/978-3-540-89707-1_5
3. Backenköhler, M., Bortolussi, L., Wolf, V.: Generalized method of moments for stochastic reaction networks in equilibrium. In: Bartocci, E., Lio, P., Paoletti, N. (eds.) CMSB 2016. LNCS, vol. 9859, pp. 15–29. Springer, Cham (2016). https://doi.org/10.1007/978-3-319-45177-0_2
4. Backenköhler, M., Bortolussi, L., Wolf, V.: Moment-based parameter estimation for stochastic reaction networks in equilibrium. IEEE/ACM Trans. Comput. Biol. Bioinf. **15**(4), 1180–1192 (2018)
5. Bartocci, E., Bortolussi, L., Nenzi, L., Sanguinetti, G.: System design of stochastic models using robustness of temporal properties. Theor. Comput. Sci. **587**, 3–25 (2015)

6. Bortolussi, L., Cardelli, L., Kwiatkowska, M., Laurenti, L.: Approximation of probabilistic reachability for chemical reaction networks using the linear noise approximation. In: Agha, G., Van Houdt, B. (eds.) QEST 2016. LNCS, vol. 9826, pp. 72–88. Springer, Cham (2016). https://doi.org/10.1007/978-3-319-43425-4_5

7. Bortolussi, L., Hillston, J., Latella, D., Massink, M.: Continuous approximation of collective system behaviour: a tutorial. Perform. Eval. **70**(5), 317–349 (2013)

8. Bortolussi, L., Sanguinetti, G.: Learning and designing stochastic processes from logical constraints. In: Joshi, K., Siegle, M., Stoelinga, M., D'Argenio, P.R. (eds.) QEST 2013. LNCS, vol. 8054, pp. 89–105. Springer, Heidelberg (2013). https://doi.org/10.1007/978-3-642-40196-1_7

9. Bortolussi, L., Silvetti, S.: Bayesian statistical parameter synthesis for linear temporal properties of stochastic models. In: Beyer, D., Huisman, M. (eds.) TACAS 2018. LNCS, vol. 10806, pp. 396–413. Springer, Cham (2018). https://doi.org/10.1007/978-3-319-89963-3_23

10. Brim, L., Češka, M., Dražan, S., Šafránek, D.: Exploring parameter space of stochastic biochemical systems using quantitative model checking. In: Sharygina, N., Veith, H. (eds.) CAV 2013. LNCS, vol. 8044, pp. 107–123. Springer, Heidelberg (2013). https://doi.org/10.1007/978-3-642-39799-8_7

11. Češka, M., Dannenberg, F., Paoletti, N., Kwiatkowska, M., Brim, L.: Precise parameter synthesis for stochastic biochemical systems. Acta Informatica **54**(6), 589–623 (2017)

12. Daca, P., Henzinger, T.A., Křetínský, J., Petrov, T.: Faster statistical model checking for unbounded temporal properties. ACM Trans. Comput. Log. (TOCL) **18**(2), 12 (2017)

13. Daws, C.: Symbolic and parametric model checking of discrete-time Markov Chains. In: Liu, Z., Araki, K. (eds.) ICTAC 2004. LNCS, vol. 3407, pp. 280–294. Springer, Heidelberg (2005). https://doi.org/10.1007/978-3-540-31862-0_21

14. de Moura, L., Bjørner, N.: Z3: an efficient SMT solver. In: Ramakrishnan, C.R., Rehof, J. (eds.) TACAS 2008. LNCS, vol. 4963, pp. 337–340. Springer, Heidelberg (2008). https://doi.org/10.1007/978-3-540-78800-3_24

15. Dehnert, C., et al.: PROPhESY: a PRObabilistic ParamEter SYnthesis tool. In: Kroening, D., Păsăreanu, C.S. (eds.) CAV 2015. LNCS, vol. 9206, pp. 214–231. Springer, Cham (2015). https://doi.org/10.1007/978-3-319-21690-4_13

16. Dehnert, C., Junges, S., Katoen, J.-P., Volk, M.: A STORM is coming: a modern probabilistic model checker. In: Majumdar, R., Kunčak, V. (eds.) CAV 2017. LNCS, vol. 10427, pp. 592–600. Springer, Cham (2017). https://doi.org/10.1007/978-3-319-63390-9_31

17. Dorigo, M., Birattari, M., Blum, C., Clerc, M., Stützle, T., Winfield, A.: Ant Colony Optimization and Swarm Intelligence, vol. 5217. Springer, Heidelberg (2008)

18. Kluyver, T., et al.: Jupyter notebooks - a publishing format for reproducible computational workflows. In: Positioning and Power in Academic Publishing: Players, Agents and Agendas, pp. 87–90. IOS Press (2016)

19. Giacobbe, M., Guet, C.C., Gupta, A., Henzinger, T.A., Paixão, T., Petrov, T.: Model checking the evolution of gene regulatory networks. Acta Informatica **54**(8), 765–787 (2017)

20. Giardina, I.: Collective behavior in animal groups: theoretical models and empirical studies. HFSP J. **2**(4), 205–219 (2008)

21. Hansen, L.P.: Large sample properties of generalized method of moments estimators. Econometrica **50**, 1029–1054 (1982)

22. Hillston, J.: Challenges for quantitative analysis of collective adaptive systems. In: Abadi, M., Lluch Lafuente, A. (eds.) TGC 2013. LNCS, vol. 8358, pp. 14–21. Springer, Cham (2014). https://doi.org/10.1007/978-3-319-05119-2_2

23. Jansen, N., et al.: Accelerating parametric probabilistic verification. In: Norman, G., Sanders, W. (eds.) QEST 2014. LNCS, vol. 8657, pp. 404–420. Springer, Cham (2014). https://doi.org/10.1007/978-3-319-10696-0_31

24. Katoen, J.-P.: The probabilistic model checking landscape. In: Proceedings of the 31st Annual ACM/IEEE Symposium on Logic in Computer Science, pp. 31–45. ACM (2016)

25. Kwiatkowska, M., Norman, G., Parker, D.: PRISM 4.0: verification of probabilistic real-time systems. In: Gopalakrishnan, G., Qadeer, S. (eds.) CAV 2011. LNCS, vol. 6806, pp. 585–591. Springer, Heidelberg (2011). https://doi.org/10.1007/978-3-642-22110-1_47

26. Loreti, M., Hillston, J.: Modelling and analysis of collective adaptive systems with CARMA and its tools. In: Bernardo, M., De Nicola, R., Hillston, J. (eds.) SFM 2016. LNCS, vol. 9700, pp. 83–119. Springer, Cham (2016). https://doi.org/10.1007/978-3-319-34096-8_4

27. Mai, M., et al.: Monitoring pre-seismic activity changes in a domestic animal collective in central Italy. In: EGU General Assembly Conference Abstracts, vol. 20, p. 19348 (2018)

28. Nouvian, M., Reinhard, J., Giurfa, M.: The defensive response of the honeybee Apis mellifera. J. Exp. Biol. **219**(22), 3505–3517 (2016)

29. Daca, P., Henzinger, T.A., Křetínský, J., Petrov, T.: Faster statistical model checking for unbounded temporal properties. In: Chechik, M., Raskin, J.-F. (eds.) TACAS 2016. LNCS, vol. 9636, pp. 112–129. Springer, Heidelberg (2016). https://doi.org/10.1007/978-3-662-49674-9_7

30. Polgreen, E., Wijesuriya, V.B., Haesaert, S., Abate, A.: Data-efficient Bayesian verification of parametric Markov Chains. In: Agha, G., Van Houdt, B. (eds.) QEST 2016. LNCS, vol. 9826, pp. 35–51. Springer, Cham (2016). https://doi.org/10.1007/978-3-319-43425-4_3

31. Quatmann, T., Dehnert, C., Jansen, N., Junges, S., Katoen, J.-P.: Parameter synthesis for Markov models: faster than ever. In: Artho, C., Legay, A., Peled, D. (eds.) ATVA 2016. LNCS, vol. 9938, pp. 50–67. Springer, Cham (2016). https://doi.org/10.1007/978-3-319-46520-3_4

32. Schnoerr, D., Sanguinetti, G., Grima, R.: Approximation and inference methods for stochastic biochemical Kinetics–a tutorial review. J. Phys. A: Math. Theor. **50**(9), 093001 (2017)

33. Shorter, J.R., Rueppell, O.: A review on self-destructive defense behaviors in social insects. Insectes Soc. **59**(1), 1–10 (2012)

34. Sokolova, A., de Vink, E.P.: Probabilistic automata: system types, parallel composition and comparison. In: Baier, C., Haverkort, B.R., Hermanns, H., Katoen, J.-P., Siegle, M. (eds.) Validation of Stochastic Systems. LNCS, vol. 2925, pp. 1–43. Springer, Heidelberg (2004). https://doi.org/10.1007/978-3-540-24611-4_1

35. Stoelinga, M.: An introduction to probabilistic automata. Bull. EATCS **78**(176–198), 2 (2002)

36. Česka, M., Šafránek, D., Dražan, S., Brim, L.: Robustness analysis of stochastic biochemical systems. PLoS ONE **9**(4), 1–23 (2014)

37. Wu, S.-H., Smolka, S.A., Stark, E.W.: Composition and behaviors of probabilistic I/O automata. Theor. Comput. Sci. **176**(1–2), 1–38 (1997)

rPrism – A Software for Reactive Weighted State Transition Models

Daniel Figueiredo[1($\boxtimes$)], Eugénio Rocha[1], Manuel António Martins[1],
and Madalena Chaves[2]

[1] CIDMA – University of Aveiro, Aveiro, Portugal
daniel.figueiredo@ua.pt
[2] Inria – Sophia Antipolis, Mediterranée, Valbonne, France

Abstract. In this work we introduce the software rPrism, as a branch of the software *PRISM model checker*, in order to be able to study weighted reactive state transition models. This kind of model gathers together the concepts of reactivity – which consists of the capacity of a state transition model to alter its accessibility relation – and weights, which can be seen as costs, rates, *etc.*. Given a specific model, the tool performs a simulation based on a Continuous Time Markov Chain. In particular, we show an example of its application for biological systems.

Keywords: *rPrism* · *PRISM model checker* · Reactive models · Weighted switch graphs

1 Introduction

The concept of reactivity in state transition models have been introduced by several authors such as van Benthem, Areces and Gabbay and some examples can be found in [1–3,8,9]. These reactive models are those whose accessibility relation (set of edges) is not fixed but can vary according to a taken path. In some sense, it can be seen as a model with memory.

The authors mentioned before, proposed several formalisms to study such models. In this paper, we will focus on the approach of [8]. In this paper, switch graphs are presented and their application is illustrated with some examples. In particular, systems whose dynamics can be described using counters or demanding some specific order to evolve are shown to be more efficiently described by reactive models. Also in biology, this kind of model can be applied: a previous work in a related topic, where a reactive model is proposed for the study of biological regulatory networks can be found in [6].

Here, we present the tool rPrism that was designed as a branch of PRISM model checker [10] to study such reactive models. The proposed tool calls PRISM to simulate the evolution of a reactive state transition model.

© Springer Nature Switzerland AG 2019
M. Češka and N. Paoletti (Eds.): HSB 2019, LNBI 11705, pp. 165–174, 2019.
https://doi.org/10.1007/978-3-030-28042-0_11

1.1   Background

We start by introducing some theoretical foundations in the topic in order to be able to explain better the relevance and usability of rPrism.

Definition 1 (Switch graph). *A switch graph is a pair (W, S) such that W is a non empty set of states and S is defined recursively as:*

- $S_0 \subseteq W \times W$;
- $S_{i+1} \subseteq S_0 \times S_i \times \{\circ, \bullet\}$, *for* $i \in \mathbb{Z}_0^+$;
- $S = \bigcup_{i \in \mathbb{Z}_0^+} S_i$.

We say that S is the set of edges and the edges in $S \backslash S_0$ are higher-level edges. Furthermore, if $e \in S_i$, e is said to be a i-level edge. Also, the initial configuration of a switch graph is given by an initial instantiation function $I_0 : W \to \{0, 1\}$.

A higher-level edge $(d, e, *)$ is said to be an activator if $*$ is $\bullet$ and is said to be an inhibitor if $*$ is $\circ$. It means that it will either inhibit (temporarily remove) or activate (reintroduce) its target edge e in the model whenever the source edge d is crossed. If the state of the target edge already agrees with the directed by the higher-level edge $(d, e, *)$, then it has no effect.

In the graphical representation of a switch graph, as shown in Fig. 1, inhibitor edges are depicted as white headed arrows, while black headed arrows represent activator edges.

At any time, the configuration of a switch graph is given through an *instantiation* function $I : S \to \{0, 1\}$ which marks each edge as inhibited (temporarily removed) or active depending on $I(s)$. An edge is active if $I(s) = 1$ and it is inhibited otherwise. The former edges are depicted as dashed arrows while the later edges are depicted as full arrows. Inhibited edges cannot be crossed, and they can neither activate nor inhibit other edges. Moreover, only 0-level edges (going from nodes to nodes) can be crossed: if one such edge x is crossed, all active higher-level edges with source in x, i.e. $(x, e, *)$ will fire and activate/inhibit the respective target edge e.

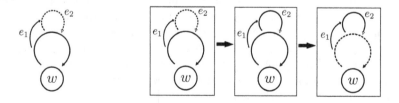

Fig. 1. Example and evolution of a switch graph.

Example 1. Figure 1 depicts a switch graph (W, S) with $W = \{w\}$ and

$$S = \{(w, w), \big((w, w), (w, w), \circ\big), \big((w, w), \big((w, w), (w, w), \circ\big)\big), \bullet\big)\}$$

For simplicity, we define $e_1 = \big((w, w), \big((w, w), \circ\big)\big), \bullet\big)$ and $e_2 = \big((w, w), (w, w), \circ\big)$. The initial instantiation I_0 is such that $I_0(w, w) = 1$ (the edge (w, w) can be crossed), $I_0(e_2) = 0$ (meaning that it is inhibited) and $I_0(e_1) = 1$ (therefore, activated and ready to activate e_2, the pointed edge whenever (w, w) is crossed). Therefore, starting from w, the edge (w, w) can be crossed (since it is active) and this causes the higher-level edge e_1 to fire and activate e_2. But e_2 has no effect since it was originally inhibited when (w, w) was crossed. One can then cross (w, w) again. Now, e_1 acts but has no effect, since e_2 is already active, while e_2 acts and inhibits (w, w). Hence, (w, w) can no longer be crossed. This is illustrated in Fig. 1.

Switch graphs can indeed be used in several fields. See [6,7] for more examples.

1.2   Weighted Switch Graphs

In this section, we introduced a generalization of switch graphs to include weights. Weights in state transition models are very useful and can represent diverse mechanisms such as costs, distances, rewards, probabilities, besides others.

Definition 2. *A weighted switch graph is a pair (W, S) together with an initial instantiation $I_0 : S \to \Omega \cup \{\circledcirc\}$ where Ω is the set of weights, and can be chosen according to the context.*

In weighted switch graphs, instead of simply considering that an edge is active or inhibited, each domain has a weight. In order to express this, we can generalize the notion of instantiation. This is attained by considering an instantiation as a function whose images belong to a set of weights Ω, along with an element $\circledcirc$ as image. Thus, if s is an edge of the model and I an instantiation, $I(s) = \circledcirc$ mean that the edge s is inhibited (temporarily removed from the model) and, otherwise, we say that the edge s is active and with weight $I(s)$.

Given this definition, we can describe the evolution of a weighted switch graph when some edge $s \in W \times W$ is crossed in the following way:

$$I^+(t) = \begin{cases} I(t), & \text{if } (s, t, *) \notin S \vee I\big((s, t, *)\big) = \circledcirc, \text{ for any } * \in \{\bullet, \circ\} \\ \circledcirc, & \text{if } (s, t, \circ) \in S \text{ and } I\big((s, t, \circ)\big) \neq \circledcirc \\ I\big((s, t, \bullet)\big), & \text{otherwise.} \end{cases}$$

Although we introduced this general definition, we only consider a particular class of models for now. The current version of the package rPrism, version 1.0, is suitable for one-level weighted switch graphs, which are defined as follows.

Definition 3. *A one-level weighted switch graph is a pair* (W, S) *with* $W \neq \{\}$ *and* $S = S_0 \cup S_1$ *such that:*

- $S_0 \subseteq W \times W$
- $S_1 \subseteq S_0 \times S_0$

along with an initial instantiation $I_0 : S \to \Omega$, *where* Ω *is the set of weights.*

Note. In this context, $\odot \equiv 0$, because our weights will be conceived as rates.

2   About the Tool rPrism

As mentioned before, rPrism is suitable to deal with one-level weighted switch graph. Furthermore, the codomain for the considered instantiations must be $\mathbb{Q}_0^+$ and, in practical cases, weights should be considered as rates. For instance, if an edge from a node w to a node w' has weight a, then it means that the component represented by w will become the component w' with rate a. The tool then performs a stochastic simulation based on a Continuous Time Markov Chain.

Given a one-level weighted switch graph, the user of rPrism must specify the model in a simple text format, with the following structure:

```
NS {
 N "definition of node 1" {
   "definition of edge 1";
   "definition of edge 2";
   ...
 }
 N "definition of node 2" {
 }
 ...
}

H1 {
   "definition of one-level edge 1";
   "definition of one-level edge 2";
   ...
}

options "command1";
output  "command2";
sim cmtc;
```

The "definition of a node" has the following format:

"node label" "lbound" "ubound" "initvalue"

where "node label" is a valid string with the name/identifier of the node; "lbound" (respectively, "ubound") is an integer determining the lower (respectively, upper) bound with respect to the number of elements of type "node" on the system; and "initvalue" is a integer with the initial value of elements of type "node" on the system.

The "definition of an edge" is done in the following way:

"target node" "initial weight"

where "target node" is a valid string with the name/identifier of the target node; and "initial weight" is a float with the initial weight of the edge.

Finally, to define a one-level edge, we must use the following code:

"source edge" "target edge" "weight"

where both "source edge" and "target edge" are strings and have the format "source node":"target node"; and "weight" is a float with the weight of the one-level edge.

The entry "command1" determines the output of the program and must be filled according to the goal of the user, for example, can be "simpath 10" (meaning the simulation will make at least 10 steps) or "simtime 5.7" (the simulation will run at least until the unit of time reaches the value 5.7). The entry "command2" can be replaced as "all" in order to obtain the entire set of outputs or restricted to any combination of the commands: "odel", "simulation_results", "simulation_plot", "reachable_sets", "transition_matrix", "labels", separated by a space.

Given a one-level switch graph, rPrism translates the introduced model into PRISM language in order to use it to study reactive models. For readers who are used to PRISM syntax, the process traduces nodes to variables and edges to actions. However, an additional module for rates is considered. There, higher-level edges are encoded as an additional variable whose value determines the rates of target edges (*actions*).

Finally, we point out that the rPrism software is implemented as a sDL package[1], and an online demo is available[2] for testing purposes. Nevertheless, the online demo has several limitations due to the fact of being over a web browser, such specific limitations do not exist when using directly the sDL client.

3 Modeling Biochemical Systems

Switch graphs can describe diverse dynamics of systems which regular graphs can not. An example is the possibility of describing counters such as the one illustrated at Fig. 1 and many others can be found in [3,7–9]. Also in biochemical contexts, we can find systems which can intuitively be described by switch

---

[1] http://sdl.mathdir.org.

[2] http://sdl-vm2.mathdir.org/demos/sDL-pck-run?pck=rPrism/1.0&sdoc=
Example_A.

graphs. Indeed, as mentioned before, an application of switch graph to the study of biological regulatory networks can be found in [6].

A simple example of a biochemical process which can be modeled using a switch graph is presented in Fig. 2: the scheme represents the general sequence of stages related to a vaccination process. Upon vaccination, a susceptible individual immediately has a lower probability of becoming infected if it comes into contact with a virus. This fact is described by the inhibition arrow.

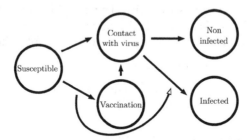

Fig. 2. Example of a switch graph describing a biological systems.

Another biological system which could be described by reactive formalisms is the *cooperativity of a hemoglobin protein* and it is described in [5]. In this system, a hemoglobin protein can bind up to 4 oxygen molecules. Thus, although this seems a simple systems, it needs to be described by a model which accommodates features such as counters, and switch graphs are perfect for the case. However, at this point, we must note that even switch graphs do not fully describe the dynamics of the mentioned system. Indeed, the cooperativity of hemoglobin is characterized by the fact that binding to one oxygen molecule increases the likelihood of binding to another one (up to the maximum of four). This increase of the binding rate cannot be described by simple switch graphs which do not consider any kind of quantitative measure. Thus, this issue is solved with weighted switch graphs which admit weights in edges.

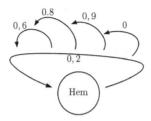

Fig. 3. Model representing the cooperativity of hemoglobin.

The example of a weighted switch graph describing a hemoglobin protein is shown in Fig. 3. There, each loop represents the binding of one oxygen molecule

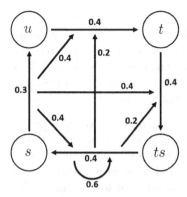

Fig. 4. Weighted switch graph of the biological circadian rhythm model.

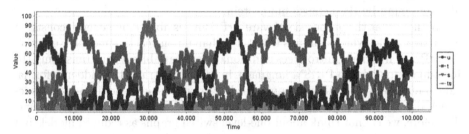

Fig. 5. Plot of the simulation.

and the weight of the loop represents the respective rate: note that the weights increase for the binding of successive oxygen molecules, but a fifth molecule can no longer bind.

Finally, we introduce another example and show how rPrism can be used to study weighted switch graph models.

Example 2 (Circadian rhythm of cyanobacteria).

In [4] we may find a model for the circadian rhythm of a cyanobacteria which considers three phosphorylated forms of the protein KaiC (s, t and ts) and an unphosphorylated form (u). Here, we omit the occurrences of protein KaiA, as related in the model of [4], in order to show that it can be represented by the reactivity of the system. In fact, a one-level weighted switch graph for such model is presented in Fig. 4.

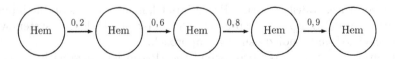

Fig. 6. Non-reactive weighted model for hemoglobin protein.

rPrism was used to simulate the evolution of this system. The plot of the output is presented in Fig. 5. Below, the code introduced in rPrism is presented.

We note that this model shows a cyclic behavior as would be expected for a circadian rhythm system. Thus, this kind of model seems suitable to represent by higher-level edges the effect of non linear or even unknown mechanisms of a cell and still obtain coherent simulations and results.

Through all examples we can find a common pattern: reactivity and, in particular, higher-level edges appear to describe the dynamics caused by a component/variable which is not considered. Indeed, note that, for instance, in the hemoglobin example we could consider a model with five states where each node is fully described by the number of oxygen molecules bound. In this way, we could obtain a model as the one shown in Fig. 6. Compared to the this one, the reactive model in Fig. 3 ignores a variable: the number of oxygen molecules already bound by the hemoglobin protein.

We note that the proposed software – rPrism – is able to, in general, construct a reactive model which contains less states than a non-reactive model. Nevertheless, rPrism internally introduces additional state variables when translating the introduced model into PRISM language, in order to retrieve the hidden information about the system. Using simple words, rPrism considers additional variables which determine what weights must be considered for each edge, at each time. Therefore, we cannot think about one-level switch graphs as reduced models but as a different description of the same model with, in general, the same "computational size". In this way, weighted switch graphs and rPrism are specially useful in two general cases:

- When the user understands that the system is more intuitively described by a reactive model, which in general depends on the background of the user.
- When the "hidden" components/variables causing reactivity are still unknown by the user.

In fact, the second point described above occurs frequently in biological contexts when, for instance, there is a missing or misunderstood regulation between two components in a system. Reactive formalisms allows one to, even so, recover coherent results from such model.

4   Conclusions and Future Work

In this paper we briefly introduce the software rPrism to study one-level weighted switch graphs. The proposed software has a proper syntax in order to be a friendly software, translating the rPrism input language to a suitable input for simulations in PRISM. Also, we present an example of a biological system which can be modeled and studied using this approach.

As future work, we intend to extend the rPrism language with the aim of exploring as much as possible the rich set of features of PRISM, namely, adding some model checking capabilities for reactive graphs. In fact, we choose to use PRISM to be the basis of our work based on its model checking capacities.

It has a temporal logic language embedded and it would allow us to compute the validity of properties as well of probabilities for some events to occur, which would be quite relevant for model analysis and predictions for biological systems. Also, we intend to expand the class of models that are suitable to be studied using rPrism. In particular, we intend to consider the general set of higher-level edges. Finally, as ongoing work, we are applying rPrism to a wider number of biological problems.

Acknowledgments. This work was supported by ERDF - The European Regional Development Fund through the Operational Programme for Competitiveness and Internationalisation - COMPETE 2020 Programme and by National Funds through the Portuguese funding agency, FCT - Fundação para a Ciência e a Tecnologia, within project POCI-01-0145-FEDER-030947 and project with reference UID/MAT/04106/2019 at CIDMA. The authors acknowledge the support given by a France-Portugal partnership PHC PESSOA 2018 between M. Chaves (Campus France #40823SD) and M. A. Martins. D. Figueiredo also acknowledges the support given by FCT via the PhD scholarship PD/BD/114186/2016.

A Appendix

The code used in rPrism for the circadian rhythm reactive model is presented bellow:

```
NS {
N s   0 100 25 {
u   0.3;
}
N ts 0 100 25 {
s   0.4;
}
N t   0 100 25 {
ts 0.4;
}
N u   0 100 25 {
t   0.4;
}
}

H1 {
s:u   u:t   0.4;
s:u   ts:s 0.4;
s:u   t:ts 0.4;
ts:s u:t   0.2;
ts:s t:ts 0.2;
ts:s ts:s 0.6;
}

options simtime 100000;
output all;
sim cmtc;
```

References

1. Areces, C., Fervari, R., Hoffmann, G.: Swap logic. Log. J. IGPL **22**, 309–332 (2013). https://doi.org/10.1093/jigpal/jzt030
2. Areces, C., Fervari, R., Hoffmann, G.: Relation-changing modal operators. Log. J. IGPL **23**, 601–627 (2015). https://doi.org/10.1093/jigpal/jzv020
3. van Benthem, J.: An essay on sabotage and obstruction. In: Hutter, D., Stephan, W. (eds.) Mechanizing Mathematical Reasoning. LNCS (LNAI), vol. 2605, pp. 268–276. Springer, Heidelberg (2005). https://doi.org/10.1007/978-3-540-32254-2_16
4. Chaves, M., Preto, M.: Hierarchy of models: from qualitative to quantitative analysis of circadian rhythms in cyanobacteria. Chaos: Interdisc. J. Nonlinear Sci. **23**(2), 025113 (2013)
5. Chou, K.C.: Low-frequency resonance and cooperativity of hemoglobin. Trends Biochem. Sci. **14**(6), 212 (1989)
6. Figueiredo, D., Barbosa, L.S.: Reactive models for biological regulatory networks. In: Chaves, M., Martins, M.A. (eds.) MLCSB 2018. LNCS, vol. 11415, pp. 74–88. Springer, Cham (2019). https://doi.org/10.1007/978-3-030-19432-1_5
7. Figueiredo, D., Martins, M.A., Barbosa, L.S.: A note on reactive transitions and Reo connectors. In: de Boer, F., Bonsangue, M., Rutten, J. (eds.) It's All About Coordination. LNCS, vol. 10865, pp. 57–67. Springer, Cham (2018). https://doi.org/10.1007/978-3-319-90089-6_4
8. Gabbay, D., Marcelino, S.: Global view on reactivity: switch graphs and their logics. Ann. Math. Artif. Intell. **66**(1–4), 131–162 (2012)
9. Gabbay, D.M., Marcelino, S.: Modal logics of reactive frames. Stud. Logica. **93**(2), 405–446 (2009)
10. Kwiatkowska, M., Norman, G., Parker, D.: PRISM 4.0: verification of probabilistic real-time systems. In: Gopalakrishnan, G., Qadeer, S. (eds.) CAV 2011. LNCS, vol. 6806, pp. 585–591. Springer, Heidelberg (2011). https://doi.org/10.1007/978-3-642-22110-1_47

Short Paper

Hybrid Modeling of Metabolic-Regulatory Networks (Extended Abstract)

Lin Liu and Alexander Bockmayr[✉]

Department of Mathematics and Computer Science, Freie Universität Berlin,
Arnimallee 6, 14195 Berlin, Germany
{linliu,bockmayr}@zedat.fu-berlin.de

Abstract. Computational approaches in systems biology have become
a powerful tool for understanding the fundamental mechanisms of cellular metabolism and regulation. However, the interplay between the
regulatory and the metabolic system is still poorly understood. In particular, there is a need for formal mathematical frameworks that allow
analyzing metabolism together with dynamic enzyme resources and regulatory events. Here, we introduce a metabolic-regulatory network model
(MRN) that allows integrating metabolism with transcriptional regulation, macromolecule production and enzyme resources. Using this model,
we show that the dynamic interplay between these different cellular
processes can be formalized by a hybrid system, combining continuous
dynamics and discrete control.

Keywords: Computational modeling · Metabolism ·
Resource allocation · Gene regulation · Hybrid system

1 Introduction

Computational approaches in systems biology have become a powerful tool for
understanding the fundamental mechanisms of cellular metabolism and regulation. Concerning integrated modeling of metabolism and regulation, there
exist approaches such as [1,2] that combine Boolean or multi-valued logical
rules for transcriptional regulation with a steady-state stoichiometric model of
metabolism. These techniques iterate flux balance analysis (FBA) by splitting
the growth phase into discrete time steps. At each time step, the updated regulatory states are imposed as bounds on the reaction fluxes while ignoring the
costs for enzyme production. At a different level, there exist methods to predict
metabolic resource allocation considering enzyme-catalytic relationships, either
at steady-state [3,4] or in a dynamic setting [5–7]. But, regulation is not included
in these approaches. Besides Boolean logic and stoichiometric models, piecewise-
linear differential equations [8,9] and other types of hybrid systems [10,11] have
also been used to study the dynamics of metabolic-genetic networks. Most of
these studies, however, merely consider metabolism and regulation, and do not
combine these with macromolecule production and enzymatic relationships.

© Springer Nature Switzerland AG 2019
M. Česka and N. Paoletti (Eds.): HSB 2019, LNBI 11705, pp. 177–180, 2019.
https://doi.org/10.1007/978-3-030-28042-0_12

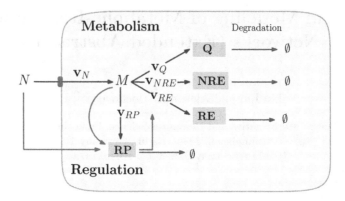

Fig. 1. Schematic model of the metabolic-regulatory network (MRN).

In the present work, we introduce a metabolic-regulatory network model (MRN) extending the self-replicator system proposed in [12]. Our modeling framework allows integrating metabolism with transcriptional regulation, macromolecule production, enzyme resources, and structural building blocks. Using this framework, we show that the dynamic interplay between cellular metabolism, macromolecule production and regulation can be formalized by a hybrid system, combining continuous dynamics and discrete control. In this formalization, all metabolite concentrations are represented by continuous variables. The discrete states of the system are composed of all gene expression states for the regulated proteins, which include regulatory proteins and regulated enzymes. In each discrete state, the continuous variables evolve according to a system of differential equations that is specific for this state. The guard conditions for the state transitions depend on the concentrations of the molecular species and associated thresholds.

Our formalization makes it possible to apply hybrid system tools for analyzing metabolic-regulatory cellular processes. Compared to the approaches mentioned above, this will allow us including regulation, macromolecule production and enzyme resources into the prediction of the dynamics of cellular metabolism.

2   Metabolic-Regulatory Networks

We formalize the interactions between metabolism and regulation by a metabolic-regulatory network (MRN) that is given in Fig. 1. Regarding metabolism, $\mathbf{N}$ represents the set of external nutrients and $\mathbf{v_N}$ is the set of intermediate reactions that convert the nutrients into precursor metabolites $\mathbf{M}$. The macromolecular production reactions $\mathbf{v_{RP}}, \mathbf{v_Q}$ and $\mathbf{v_E} = \mathbf{v_{RE}} \cup \mathbf{v_{NRE}}$ use the precursors $\mathbf{M}$ to build regulatory proteins $\mathbf{RP}$, non-catalytic macromolecules $\mathbf{Q}$, and enzymes $\mathbf{E}$. To keep the model simple, the set of enzymes $\mathbf{E}$ contains all catalytic molecules, including transporters and ribosomes. However, we distinguish between regulated enzymes $\mathbf{RE}$ and non-regulated enzymes $\mathbf{NRE}$, i.e.,

$\mathbf{E} = \mathbf{RE} \cup \mathbf{NRE}$. Non-catalytic macromolecules, termed as quota compounds $\mathbf{Q}$, e.g. DNA and lipids, are included in the model because they are essential for growth and consume a lot of cellular resources.

3 Hybrid Discrete-Continuous Dynamics

Based on our metabolic-regulatory network (see Fig. 1), we define the set of molecular species $\mathcal{M} = \mathbf{N} \cup \mathbf{M} \cup \mathbf{RP} \cup \mathbf{E} \cup \mathbf{Q}$. In a purely continuous modeling approach, the dynamics of the network would be described by a system of ordinary differential equations

$$\dot{\mathcal{M}}(t) = \frac{d\mathcal{M}}{dt} = F(\mathcal{M}, \mathbf{K}, \mathbf{S}, t), \tag{1}$$

where $\mathbf{K}$ is the set of kinetic parameters, $\mathbf{S}$ is the stoichiometric matrix, and t denotes time. The function F represents the kinetic laws that govern the dynamics, which - depending on the molecular species - could be mass action, Michaelis-Menten, Hill kinetics etc.

Continuous modeling of gene regulatory networks is known to be very difficult due to the lack of the necessary kinetic data. Therefore, we adopt a more qualitative approach to include regulation in our model. It is based on the logical modeling framework pioneered in the 1970's by L. Glass, S. Kauffman, R. Thomas et al., see [13] for a recent review. We assume that for each regulated protein p there are two possible states $\mathtt{on}$ and $\mathtt{off}$, describing whether at a particular time t the gene encoding p is expressed or not.

Formally, for all $p \in \mathbf{RP} \cup \mathbf{RE}$, we introduce a Boolean variable $\bar{p} = \bar{p}(t) \in \{0, 1\}$ and a logical function $f_p : \mathbb{R}^n \to \{0, 1\}$. Here, the Boolean value 0 corresponds to $\mathtt{off}$ and the value 1 to $\mathtt{on}$. Each function f_p is defined as a Boolean combination (using the Boolean operations $\neg$ (not), $\wedge$ (and), $\vee$ (or)) of atomic formulae $x_i \geq \theta_i$, where x_i is a real variable and θ_i is a constant. Overall, the regulation of our MRN is then formalized by a system of Boolean equations of the form

$$\bar{p}(t) = f_p(\mathbf{RP}(t), \mathbf{N}(t), \mathbf{M}(t)), \text{ for all } p \in \mathbf{RP} \cup \mathbf{RE}. \tag{2}$$

Here, f_p describes how the expression state of the gene encoding the regulated protein p depends on the current concentrations of regulatory proteins, external nutrients, and intermediate metabolites.

Combining metabolism and regulation in this way leads to a hybrid discrete-continuous system. Here, all concentrations of molecular species are modeled by continuous variables. However, the evolution of the regulated proteins p is controlled by the expression state $\bar{p}$ of the corresponding genes. Thus, depending on the discrete state $\bar{p}$, there are two different continuous dynamics. The system will jump from one discrete state to the other if some regulatory event given by Eq. (2) occurs, see Fig. 2 for illustration.

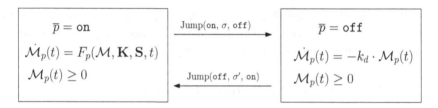

Fig. 2. Graphical representation of the continuous evolution in the discrete states $\bar{p} =$ on and $\bar{p} = \texttt{off}$, for regulated proteins $p \in \textbf{RP} \cup \textbf{RE}$. In the on-state, protein production and degradation is described by the kinetic law $F_p(\cdot)$, while in the off-state only degradation occurs, with kinetic constant k_d.

Acknowledgment. Lin Liu acknowledges support from the China Scholarship Council (CSC).

References

1. Covert, M.W., Schilling, C.H., Palsson, B.Ø.: Regulation of gene expression in flux balance models of metabolism. J. Theor. Biol. **213**(1), 73–88 (2001)
2. Marmiesse, L., Peyraud, R., Cottret, L.: FlexFlux: combining metabolic flux and regulatory network analyses. BMC Syst. Biol. **9**(1), 93 (2015)
3. Goelzer, A., Fromion, V., Scorletti, G.: Cell design in bacteria as a convex optimization problem. Automatica **47**(6), 1210–1218 (2011)
4. Lerman, J.A., Hyduke, D.R., Latif, H., Portnoy, V.A., et al.: In silico method for modelling metabolism and gene product expression at genome scale. Nat. Commun. **3**, 929 (2012)
5. Waldherr, S., Oyarzún, D.A., Bockmayr, A.: Dynamic optimization of metabolic networks coupled with gene expression. J. Theor. Biol. **365**, 469–485 (2015)
6. Rügen, M., Bockmayr, A., Steuer, R.: Elucidating temporal resource allocation and diurnal dynamics in phototrophic metabolism using conditional FBA. Sci. Rep. **5**, 15247 (2015)
7. Lloyd, C.J., et al.: COBRAme: a computational framework for genome-scale models of metabolism and gene expression. PLoS Comput. Biol. **14**(7), e1006302 (2018)
8. Ropers, D., De Jong, H., Page, M., Schneider, D., Geiselmann, J.: Qualitative simulation of the carbon starvation response in Escherichia coli. Biosystems **84**(2), 124–152 (2006)
9. Chaves, M., Oyarzún, D.A., Gouzé, J.L.: Analysis of a genetic-metabolic oscillator with piecewise linear models. J. Theor. Biol. **462**, 259–269 (2019)
10. Bockmayr, A., Courtois, A.: Using hybrid concurrent constraint programming to model dynamic biological systems. In: Stuckey, P.J. (ed.) ICLP 2002. LNCS, vol. 2401, pp. 85–99. Springer, Heidelberg (2002). https://doi.org/10.1007/3-540-45619-8_7
11. Bortolussi, L., Policriti, A.: Hybrid systems and biology. In: Bernardo, M., Degano, P., Zavattaro, G. (eds.) SFM 2008. LNCS, vol. 5016, pp. 424–448. Springer, Heidelberg (2008). https://doi.org/10.1007/978-3-540-68894-5_12
12. Molenaar, D., Van Berlo, R., De Ridder, D., Teusink, B.: Shifts in growth strategies reflect tradeoffs in cellular economics. Mol. Syst. Biol. **5**(1), 323 (2009)
13. Abou-Jaoudé, W., et al.: Logical modeling and dynamical analysis of cellular networks. Front. Genet. **7**, 94 (2016)

Author Index

Printed in the United States
By Bookmasters